T0334249

Artificial Cognitive Systems
A Primer

Artificial Cognitive Systems
A Primer

David Vernon

The MIT Press
Cambridge, Massachusetts
London, England

This book was set in Palatino by the author.

Library of Congress Cataloging-in-Publication Data

Vernon, David, 1958–
Artificial cognitive systems : a primer / David Vernon.
    p.   cm
Includes bibliographical references and index.
ISBN 978-0-262-02838-7 (hardcover : alk. paper)
ISBN 978-0-262-55287-5 (pb)
1. Cognition. 2. Brain. 3. Development psychology. 4. Artificial intelligence.   I. Title.
BF311.V447   2014
153—dc23

2014013208

To Keelin, Ciana, and Georgina —
for never giving up on your dreams

"It's a poor sort of memory that only works backwards."

Lewis Carroll — *Through the Looking Glass*

.

Contents

Preface

This primer introduces you to the emerging field of artificial cognitive systems. Inspired by artificial intelligence, developmental psychology, and cognitive neuroscience, the aim is to build systems that can act on their own to achieve goals: perceiving their environment, anticipating the need to act, learning from experience, and adapting to changing circumstances.

It is an exciting and challenging area. The excitement stems from the possibility of designing intelligent adaptive systems that can serve society in a host of ways. The challenge is the breadth of the field and the need to bring together an intimidating spectrum of disciplines. Add to this the fact that there is no universal agreement on what exactly it means to be cognitive in the first place and the stage is set for an interesting journey. Think of this primer as a guidebook to help you on that journey, pointing out the main features of the landscape, the principal routes, and the most important landmarks.

To get started, we develop a working definition of cognitive systems, one that strikes a balance between being broad enough to do service to the many views that people have on cognition and deep enough to help in the formulation of theories and models. We then survey the different paradigms of cognitive science to establish the full scope of the subject and sketch the main geography of the area. We follow that with a discussion of cognitive architectures — effectively the blueprints for implementing cognitive systems — before tackling the key issues, one by one, in the remaining chapters: autonomy, embodiment, learning & development, memory & prospection, knowledge & representation, and social cognition.

By the time you have finished reading this primer, you will have a clear understanding of the scope of the domain, the different perspectives, and their underlying differences. You will have a solid grasp of the issues that need to be addressed when attempting to design an artificial cognitive system. Perhaps most important of all, you will know where to go next to deepen your understanding of the area and its constituent disciplines.

Like all guidebooks, this primer tells a story about the land it surveys. In fact, it tells two stories in parallel, one in the main narrative and another through a sequence of sidenotes. The main text is kept as short and simple as possible, focussing on relatively straightforward descriptions of the key issues. The sidenotes highlight the finer points of the material being discussed in the main narrative, suggesting material that you can read to gain a deeper insight into the topic under discussion. New ideas are introduced in a natural intuitive order, building step-by-step to a clear overview of this remarkable and exciting field, priming you to go further and faster in your studies of cognitive systems.

Ideally, you will read the primer three times. On the first reading, you might read only the main narrative to get a feeling for the topic. You might then read through the sidenotes without reference to the main text. This will expose you to a series of interesting snapshots of key landmark topics and reinforce ideas you encountered on the first reading. Finally, you should be ready for a third, more careful reading of the book, referring to each sidenote as it is referenced in the main narrative.

A primer, by its very nature, is a short book. Consequently, there are many omissions in this text, some intentional, others less so. By far, the topic that is most noticable by its absence is language. While providing an overview of areas such as embodiment and autonomy is a challenge because of their diverse meanings, the task of doing the same for language is far greater. So, rather than attempt it and almost inevitably fall short, I have omitted it. If there is ever a second edition, the inclusion of language will be the top priority.

Other omissions are more methodological. This primer focusses almost exclusively on the "What?" and "Why?" questions in cognitive systems, to the exclusion of the "How?" In

other words, it does everything except tell you how you can
go about building a cognitive system. There is an unfortunate,
but inevitable, lack of formal theory and algorithmic practice
in this book. This doesn't mean that this theory doesn't exist —
it certainly does, as a quick scan of the literature on, e.g., machine learning and computer vision will demonstrate — but the
breadth of cognitive systems is so great that to address the computational and mathematical theories as well as the algorithmic
and representational details of cognition would require a book of
far greater scope than this one. Perhaps, some day, a companion
volume might be appropriate.

Skövde, Sweden   *David Vernon*
May 2014

Acknowledgements

My interest in cognitive systems was ignited in 1984 by Dermot Furlong, Trinity College, Dublin, who introduced me to the seminal work of Humberto Maturana and Francisco Varela. In the intervening 30 years, he has continued to prompt, question, and debate, and his insights have helped greatly in putting the many different aspects of cognition in perspective.

Giulio Sandini, Istituto Italiano di Tecnologia (IIT), played a pivotal role in the writing of this book. Twenty years after we first worked together on an image understanding project, we collaborated again in 2004 on his brainchild, the iCub, the open-source 53 degree-of-freedom cognitive humanoid robot featured on the front cover. In this 5-year research project, funded by the European Commission, we investigated many aspects of artificial cognitive systems, and I would like to thank him sincerely for involving me in the project and for his insights and inspiration.

In 2003, Henrik Christensen, then at KTH in Sweden and now at Georgia Tech in the USA, and Hans-Hellmut Nagel, University of Karlsruhe (now Karlsruhe Institute of Technology), organized a lively workshop on cognitive vision systems at Schloss Dagstuhl. The discussions at this workshop had a strong influence in determining the content of the book and in achieving, I hope, a balanced perspective. I would like to pay a special tribute to Hans-Hellmut for his remarkable attention to detail and passion for clarity of expression. I learned much from him and I have tried to put it into practice when writing this book.

From 2002 to 2005, I coordinated *ECVision*, the EU-funded European research network for cognitive vision systems. Many of the ideas discussed at the brainstorming sessions conducted in the development of the *ECVision* research roadmap were crucial in developing my ideas on cognitive systems.

Beginning in 2006, I coordinated *euCognition*, the European Network for the Advancement of Artificial Cognitive Systems, for three years. The members — then numbering 300 or so, now over 800 — are drawn from many disciplines, including neuroscience, psychology, computer science, control, cognitive science, linguistics, cybernetics, dynamical and self-organizing systems, computer vision, and robotics. I have benefitted enormously from being exposed to their thoughts and insights, and those of our guest speakers at network meetings.

In the context of my work in European projects, I would like to say a special thank you to Horst Forster, Colette Maloney, Hans-Georg Stork, Cécile Huet, Juha Heikkilä, Franco Mastroddi, and their colleagues in the European Commission for their unstinting support. Europe's vibrant cognitive systems community is due in no small part to their foresight and leadership.

I wish to thank Gordon Cheng and Uwe Haass for giving me the opportunity to work at the Institute for Cognitive Systems (ICS), Technical University of Munich, in 2011 and 2012. My work at ICS provided the initial impetus to turn my lecture notes into a textbook and the time to do it.

Marcia Riley, a researcher at ICS, provided the intellectual flint that is essential for sparking new ideas and better ways of communicating them. I am grateful for her time, knowledge, and willingness to debate the finer points of robotics and cognition.

During my time in Munich, I had many fruitful conversations with Michael Beetz, University of Bremen. His insights gave me a fresh appreciation of the importance of new AI in cognitive systems and a much better understanding of the ways knowledge can be shared between people and robots.

If my time at the Technical University of Munich provided the impetus, my move to the University of Skövde, Sweden, in early 2013 provided the ideal environment to write the bulk of the book. Here I have the pleasure of working with great people — Tom Ziemke, Serge Thill, Paul Hemeren, Erik Billing, Rob Lowe, Jessica Lindblom, and many others — all of whom contributed directly and indirectly to the throughts expressed in the chapters that follow. Heartfelt thanks to each and every one of you.

The insights of Claes von Hoftsten, University of Uppsala, and Luciano Fadiga, University of Ferrara and Istituto Italiano di Tecnologia, contributed greatly to the material in Chapter 6. I am obliged to them both for taking the time to explain the basis of human development to me when we collaborated on the design of the iCub cognitive humanoid robot.

During my time working on the iCub, I gave several short courses on cognitive systems at the University of Genoa and the Istituto Italiano di Tecnologia. The notes for these courses formed the original basis of this book while a paper co-authored by Giulio Sandini, IIT, and Giorgio Metta, IIT, in 2007 provided the foundations for Chapter 2.

I am grateful to Alessandra Sciutti, Istituto Italiano di Tecnologia, for allowing me to use her survey data on milestones in the development of human infants in Table 6.1, Chapter 6.

The material in Chapter 9 on joint action, shared intention, and collaborative behaviour derives in part from the contributions made by Harold Bekkering, Radboud University Nijmegen, Yiannis Demiris, Imperial College, London, Giulio Sandini, Istituto Italiano di Tecnologia, and Claes von Hofsten, University of Uppsala, to a research proposal we worked on in 2012.

Thanks go to Alan Bushnell, Phoenix Technical Co. L.L.C., for taking the time to cast a cold eye on the manuscript and for his many helpful suggestions.

Books need publishers and editors, and it is a pleasure to acknowledge the part played by Philip Laughlin, Senior Acquisitions Editor at MIT Press, in bringing this particular book to completion. Without his patience and support, I wouldn't be writing these acknowledgements. Thanks too to Virginia Crossman, Christopher Eyer, Susan Clark, and Erin Hasley at MIT Press for all their help in transforming my initial draft into the finished product.

The front cover picture — featuring the iCub cognitive humanoid robot and Lorenzo Natale, IIT — is used with the kind permission of the Department of Robotics, Brain and Cognitive Sciences, Istituto Italiano di Tecnologia. The photograph was taken by Massimo Brega.

The final word of appreciation goes most especially to my wonderful wife Keelin, for patiently putting up with me working on the book during goodness-knows how many evenings, weekends, and holidays and for painstakingly proof-reading every page. Such love and understanding is exceedingly rare and I will be forever grateful.

1

The Nature of Cognition

1.1 Motivation for Studying Artificial Cognitive Systems

When we set about building a machine or writing a software application, we usually have a clear idea of what we want it to do and the environment in which it will operate. To achieve reliable performance, we need to know about the operating conditions and the user's needs so that we can cater for them in the design. Normally, this isn't a problem. For example, it is straightforward to specify the software that controls a washing machine or tells you if the ball is out in a tennis match. But what do we do when the system we are designing has to work in conditions that aren't so well-defined, where we cannot guarantee that the information about the environment is reliable, possibly because the objects the system has to deal with might behave in an awkward or complicated way, or simply because unexpected things can happen?

Let's use an example to explain what we mean. Imagine we wanted to build a robot that could help someone do the laundry: load a washing machine with clothes from a laundry basket, match the clothes to the wash cycle, add the detergent and conditioner, start the wash, take the clothes out when the wash is finished, and hang them up to dry (see Figure 1.1). In a perfect world, the robot would also iron the clothes,[1] and put them back in the wardrobe. If someone had left a phone, a wallet, or something else in a pocket, the robot should either remove it before putting the garment in the wash or put the garment to

[1] The challenge of ironing clothes as a benchmark for robotics [1] was originally set by Maria Petrou [2]. It is a difficult task because clothes are flexible and unstructured, making them difficult to manipulate, and ironing requires careful use of a heavy tool and complex visual processing.

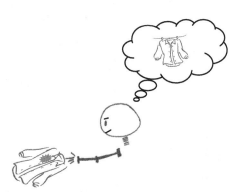

Figure 1.1: A cognitive robot would be able to see a dirty garment and figure out what needs to be done to wash and dry it.

one side to allow a human to deal with it later. This task is well beyond the capabilities of current robots[2] but it is something that humans do routinely. Why is this? It is because we have the ability to look at a situation, figure out what's needed to achieve some goal, anticipate the outcome, and take the appropriate actions, adapting them as necessary. We can determine which clothes are white (even if they are very dirty) and which are coloured, and wash them separately. Better still, we can also learn from experience and adapt our behaviour to get better at the job. If the whites are still dirty after being washed, we can apply some extra detergent and wash them again at a higher temperature. And best of all, we usually do this all on our own, autonomously, without any outside help (except maybe the first couple of times). Most people can work out how to operate a washing machine without reading the manual, we can all hang out damp clothes to dry without being told how to do it, and (almost) everyone can anticipate what will happen if you wash your smartphone.

We often refer to this human capacity for self-reliance, for being able to figure things out, for independent adaptive anticipatory action, as *cognition*. What we want is the ability to create machines and software systems with the same capacity, i.e., *artificial cognitive systems*. So, how do we do it? The first step would be to model cognition. And this first step is, unfortunately, where things get difficult because cognition means

[2] Some progress has been made recently in developing a robot that can fold clothes. For example, see the article "Cloth grasp point detection based on multiple-view geometric cues with application to robotic towel folding" by Jeremy Maitin-Shepard *et al.* [3] which describes how the PR2 robot built by Willow Garage [4] tackles the problem. However, the focus in this task is not so much the ill-defined nature of the job — how do you sort clothes into different batches for washing and, in the process, anticipate, adapt, and learn — as it is on the challenge of vision-directed manipulation of flexible materials.

different things to different people. The issue turns on two key concerns: (a) the purpose of cognition — the role it plays in humans and other species, and by extension, the role it should play in artificial systems — and (b) the mechanisms by which the cognitive system fulfils that purpose and achieves its cognitive ability. Regrettably, there's huge scope for disagreement here and one of the main goals of this book is to introduce you to the different perspectives on cognition, to explain the disagreements, and to tease out their differences. Without understanding these issues, it isn't possible to begin the challenging task of developing artificial cognitive systems. So, let's get started.

1.2   Aspects of Modelling Cognitive Systems

There are four aspects which we need to consider when modelling cognitive systems:[3] how much inspiration we take from natural systems, how faithful we try to be in copying them, how important we think the system's physical structure is, and how we separate the identification of cognitive capability from the way we eventually decide to implement it. Let's look at each of these in turn.

[3] For an alternative view that focusses on assessing the contributions made by particular models, especially computational and robotic models, see Anthony Morse's and Tom Ziemke's paper "On the role(s) of modelling in cognitive science" [5].

To replicate the cognitive capabilities we see in humans and some other species, we can either invent a completely new solution or draw inspiration from human psychology and neuroscience. Since the most powerful tools we have today are computers and sophisticated software, the first option will probably be some form of computational system. On the other hand, psychology and neuroscience reflect our understanding of biological life-forms and so we refer to the second option as a bio-inspired system. More often than not, we try to blend the two together. This balance of pure computation and bio-inspiration is the first aspect of modelling cognitive systems.

Unfortunately, there is an unavoidable complication with the bio-inspired approach: we first have to understand how the biological system works. In essence, this means we must come up with a model of the operation of the biological system and then use this model to inspire the design of the artificial system. Since biological systems are very complex, we need to choose the level

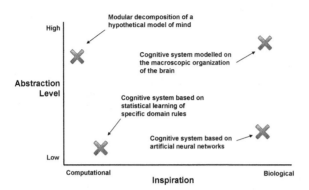

Figure 1.2: Attempts to build an artificial cognitive system can be positioned in a two-dimensional space, with one axis defining a spectrum running from purely computational techniques to techniques strongly inpired by biological models, and with another axis defining the level of abstraction of the biological model.

of abstraction at which we study them. For example, assuming for the moment that the centre of cognitive function is the brain (this might seem a very safe assumption to make but, as we'll see, there's a little more to it than this), then you might attempt to replicate cognitive capacity by emulating the brain at a very high level of abstraction, e.g. by studying the broad functions of different regions in the brain. Alternatively, you might opt for a low level of abstraction by trying to model the exact electrochemical way that the neurons in these regions actually operate. The choice of abstraction level plays an important role in any attempt to model a bio-inspired artificial cognitive system and must be made with care. That's the second aspect of modelling cognitive systems.

Taking both aspects together — bio-inspiration and level of abstraction — we can position the design of an artificial cognitive system in a two-dimensional space spanned by a computational / bio-inspired axis and an abstraction-level axis; see Figure 1.2. Most attempts today occupy a position not too far from the centre, and the trend is to move towards the biological side of the computational / bio-inspired spectrum and to cover several levels of abstraction.

In adopting a bio-inspired approach at any level of abstraction it would be a mistake to simply replicate brain mechanisms in complete isolation in an attempt to replicate cognition. Why? Because the brain and its associated cognitive capacity is the result

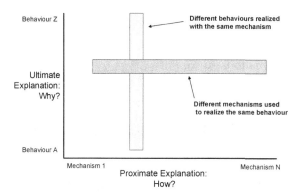

Behaviour Z

Ultimate
Explanation:
Why?

Behaviour A

Mechanism 1   Mechanism N

Proximate Explanation:
How?

Different behaviours realized
with the same mechanism

Different mechanisms used
to realize the same behaviour

Figure 1.3: The ultimate-proximate distinction. Ultimate explanations deal with *why* a given behaviour exists in a system, while proximate explanations address the specific mechanisms by which these behaviours are realized. As shown here, different mechanisms could be used to achieve the same behaviour or different behaviours might be realized with the same mechanism. What's important is to understand that identifying the behaviours you want in a cognitive system and finding suitable mechanisms to realize them are two separate issues.

of evolution and the brain evolved for some purpose. Also, the brain and the body evolved together and so you can't divorce one from the other without running the risk of missing part of the overall picture. Furthermore, this brain-body evolution took place in particular environmental circumstances so that the cognitive capacity produced by the embodied brain supports the biological system in a specific ecological niche. Thus, a complete picture may really require you to adopt a perspective that views the brain and body as a complete system that operates in a specific environmental context. While the environment may be uncertain and unknown, it almost always has some in-built regularities which are exploited by brain-body system through its cognitive capacities in the context of the body's characteristics and peculiarities. In fact, the whole purpose of cognition in a biological system is to equip it to deal with this uncertainty and the unknown nature of the system's environment. This, then, is the third aspect of modelling cognitive systems: the extent to which the brain, body, and environment depend on one another.[4]

Finally, we must address the two concerns we raised in the opening section, i.e., the purpose of cognition and the mechanisms by which the cognitive system fulfils that purpose and achieves its cognitive ability. That is, in drawing on bio-inspiration, we need to factor in two complementary issues: what cognition is for and how it is achieved. Technically, this is known as the

[4] We return to the relationship between the brain, body, and environment in Chapter 5 on embodiment.

ultimate-proximate distinction in evolutionary psychology; see Figure 1.3. Ultimate explanations deal with questions concerned with *why* a given behaviour exists in a system or is selected through evolution, while proximate explanations address the specific mechanisms by which these behaviours are realized. To build a complete picture of cognition, we must address both explanations. We must also be careful not to get the two issues mixed up, as they very often are.[5] Thus, when we want to build machines which are able to work outside known operating conditions just like humans can — to replicate the cognitive characteristics of smart people — we must remember that this smartness may have arisen for reasons other than the ones in which it is being deployed in the current task-at-hand. Our brains and bodies certainly didn't evolve so that we could load and unload a washing machine with ease, but we're able to do it nonetheless. In attempting to use bio-inspired cognitive capabilites to perform utilitarian tasks, we may well be just piggy-backing on a deeper and quite possibly quite different functional capacity. The core problem then is to ensure that this *system* functional capacity matches the ones we need to get *our* job done. Understanding this, and keeping the complementary issues of the purpose and mechanisms of cognition distinct, allows us to keep to the forefront the important issue of how one can get an artificial cognitive system (and a biological one, too, for that matter) to do what we want it to do. If we are having trouble doing this, the problem may not be the operation of the specific (proximate) mechanisms of the cognitive model but the (ultimate) selection of the cognitive behaviours and their fitness for the given purpose in the context of the brain-body-mind relationship.

To sum up, in preparing ourselves to study artificial cognitive systems, we must keep in mind four important aspects when modelling cognitive systems:

1. The computational / bio-inspired spectrum;

2. The level of abstraction in the biological model;

3. The mutual dependence of brain, body, and environment;

4. The ultimate-proximate distinction (why *vs.* how).

[5] The importance of the ultimate-proximate distinction is highlighted by Scott-Phillips *et al.* in a recent article [6]. This article also points out that ultimate and proximate explanations of phenomena are often confused with one another so we end up discussing proximate concerns when we really should be discussing ultimate ones. This is very often the case with artificial cognitive systems where there is a tendency to focus on the proximate issues of *how* cognitive mechanisms work, often neglecting the equally important issue of *what* purpose cognition is serving in the first place. These are two complementary views and both are needed. See [7] and [8] for more details on the ultimate-proximate distinction.

Understanding the importance of these four aspects will help us make sense of the different traditions in cognitive science, artificial intelligence, and cybernetics (among other disciplines) and the relative emphasis they place on the mechanisms and the purpose of cognition. More importantly, it will ensure we are addressing the right questions in the right context in our efforts to design and build artificial cognitive systems.

1.3   So, What Is Cognition Anyway?

It should be clear from what we have said so far that in asking "what is cognition?" we are posing a badly-framed question: what cognition *is* depends on what cognition is *for* and *how* cognition is realized in physical systems — the ultimate and proximate aspects of cognition, respectively. In other words, the answer to the question depends on the context — on the relationship between brain, body, and environment — and is heavily coloured by which cognitive science tradition informs that answer. We devote all of Chapter 2 to these concerns. However, before diving into a deep discussion of these issues, we'll spend a little more time here setting the scene. In particular, we'll provide a generic characterization of cognition as a preliminary answer to the question "what is cognition?", mainly to identify the principal issues at stake in designing artificial cognitive systems and always mindful of the need to explain how a given system addresses the four aspects of modelling identified above. Now, let's cut to the chase and answer the question.

Cognition implies an ability to make inferences about events in the world around you. These events include those that involve the cognitive agent itself, its actions, and the consequences of those actions. To make these inferences, it helps to remember what happened in the past since knowing about past events helps to anticipate future ones.[6] Cognition, then, involves predicting the future based on memories of the past, perceptions of the present, and in particular anticipation of the behaviour[7] of the world around you and, especially, the effects of your actions in it. Notice we say actions, not movement of motions. Actions usually involve movement or motion but an action also involves

[6] We discuss the forward-looking role of memory in anticipating events in Chapter 7.

[7] Inanimate objects don't behave but animate ones do, as do inanimate objects being controlled by animate ones (e.g. cars in traffic). So agency, direct or indirect, is implied by behaviour.

something else. This is the *goal* of the action: the desired outcome, typically some change in the world. Since predictions are rarely perfect, a cognitive system must also learn by observing what does actually happen, assimilate it into its understanding, and then adapt the way it subsequently does things. This forms a continuous cycle of self-improvement in the system's ability to anticipate future events. The cycle of anticipation, assimilation, and adaptation supports — and is supported by — an on-going process of action and perception; see Figure 1.4.

We are now ready for our preliminary definition.

> Cognition is the process by which an autonomous system perceives its environment, learns from experience, anticipates the outcome of events, acts to pursue goals, and adapts to changing circumstances.[8]

We will take this as our preliminary definition of cognition and, depending on the approach we are discussing, we will adjust it accordingly in later chapters.

While definitions are convenient, the problem with them is that they have to be continuously amended as we learn more about the thing they define.[9] So, with that in mind, we won't become too attached to the definition and we'll use it as a memory aid to remind us that cognition involved at least six attributes of autonomy, perception, learning, anticipation, action, and adaptation.

For many people, cognition is really an umbrella term that covers a collection of skills and capabilities possessed by an agent.[10] These include being able to do the following.

- Take on goals, formulate predictive strategies to achieve them, and put those strategies into effect;

- Operate with varying degrees of autonomy;

- Interact — cooperate, collaborate, communicate — with other agents;

- Read the intentions of other agents and anticipate their actions;

- Sense and interpret expected and unexpected events;

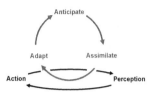

Figure 1.4: Cognition as a cycle of anticipation, assimilation, and adaptation: embedded in, contributing to, and benefitting from a continuous process of action and perception.

[8] These six attributes of cognition — autonomy, perception, learning, anticipation, action, adaptation — are taken from the author's definition of cognitive systems in the Springer *Encyclopedia of Computer Vision* [9]

[9] The Nobel laureate, Peter Medawar, has this to say about definitions: "My experience as a scientist has taught me that the comfort brought by a satisfying and well-worded definition is only short-lived, because it is certain to need modification and qualification as our experience and understanding increase; it is explanations and descriptions that are needed" [10]. Hopefully, you will find understandable explanations in the pages that follow.

[10] We frequently use the term *agent* in this book. It means any system that displays a cognitive capacity, whether it's a human, or (potentially, at least) a cognitive robot, or some other artificial cognitive entity. We will use *agent* interchangably with *artifical cognitive system*.

Figure 1.5: Another aspect of cognition: effective interaction. Here the robot anticipates someone's needs (see Chapter 9, Section 9.4 *Instrumental Helping*).

- Anticipate the need for actions and predict the outcome of its own actions and those of others;

- Select a course of action, carry it out, and then assess the outcome;

- Adapt to changing circumstances, in real-time, by adjusting current and anticipated actions;

- Learn from experience: adjust the way actions are selected and performed in the future;

- Notice when performance is degrading, identify the reason for the degradation, and take corrective action.

These capabilities focus on what the agent should do: its functional attributes. Equally important are the effectiveness and the quality of its operation: its non-functional characteristics (or, perhaps more accurately, its meta-functional characteristics): its dependability, reliability, usability, versatility, robustness, fault-tolerance, and safety, among others.[11]

These meta-functional characteristics are linked to the functional attributes through system capabilities that focus not on carrying out tasks but on maintaining the integrity of the agent.[12] Why are these capabilities relevant to artificial agents? They are relevant — and critically so — because artificial agents such as a robot that is deployed outside the carefully-configured environments typical of many factory floors have to deal with a

[11] The "non-" part of "non-functional" is misleading as it suggests a lesser value compared to functional characteristics whereas, in reality, these characteristics are equally important but complementary to functionality when designing a system. For that reason, we sometimes refer to them as meta-functional attributes; see [11] for a more extensive list and discussion of meta-functionional attributes.

[12] We will come back to the issue of maintaining integrity several times in this book, briefly in the next section, and more at length in the next chapter. For the moment, we will just remark that the processes by which integrity is maintained are known as *autonomic* processes.

world that is only partially known. It has to work with incomplete information, uncertainty, and change. The agent can only cope with this by exhibiting some degree of cognition. When you factor interaction with people into the requirements, cognition becomes even more important. Why? Because people are cognitive and they behave in a cognitive manner. Consequently, any agent that interacts with a human needs to be cognitive to some degree for that interaction to be useful or helpful. People have their own needs and goals and we would like our artificial agent to be able to anticipate these (see Figure 1.5). That's the job of cognition.

So, in summary, cognition is not to be seen as some module in the brain of a person or the software of a robot — a planning module or a reasoning module, for example — but as a system-wide process that integrates all of the capabilities of the agent to endow it with the six attributes we mentioned in our memory-aid definition: autonomy, perception, learning, anticipation, action, and adaptation.

1.3.1   Why Autonomy?

Notice that we included autonomy in our definition. We need to be careful about this. As we will see in Chapter 4, the concept of autonomy is a difficult one. It means different things to different people, ranging from the fairly innocent, such as being able to operate without too much help or assistance from others, to the more controversial, which sees cognition as one of the central processes by which advanced biological systems preserve their autonomy. From this perspective, cognitive development has two primary functions: (1) to increase the system's repertoire of effective actions, and (2) to extend the time-horizon of its ability to anticipate the need for and outcome of future actions.[13]

Without wishing to preempt the discussion in Chapter 4, because there is a tight relationship between cognition and autonomy — or not, depending on who you ask — we will pause here just a while to consider autonomy a little more.

From a biological perspective, autonomy is an organizational characteristic of living creatures that enables them to use their

[13] The increase of action capabilities and the extension anticipation capabilities as the primary focus of cognition is the central message conveyed in *A Roadmap for Cognitive Development in Humanoid Robots* [12], a multi-disciplinary book co-written by the author, Claes von Hofsten, and Luciano Fadiga.

own capacities to manage their interactions with the world in order to remain viable, i.e., to stay alive. To a very large extent, autonomy is concerned with the system maintaining itself: self-maintenance, for short.[14] This means that the system is entirely self-governing and self-regulating. It is not controlled by any outside agency and this allows it to stand apart from the rest of the environment and assert an identity of its own. That's not to say that the system isn't influenced by the world around it, but rather that these influences are brought about through interactions that must not threaten the autonomous operation of the system.[15]

If a system is autonomous, its most important goal is to preserve its autonomy. Indeed, it must act to preserve it since the world it inhabits that may not be very friendly. This is where cognition comes in. From this (biological) perspective, cognition *is* the process whereby an autonomous self-governing system acts effectively in the world in which it is embedded in order to maintain its autonomy.[16] To act effectively, the cognitive system must sense what is going on around it. However, in biological agents, the systems responsible for sensing and interpretation of sensory data, as well as those responsible for getting the motor systems ready to act, are actually quite slow and there is often a delay between when something happens and when an autonomous biological agent comprehends what has happened. This delay is called *latency* and it is often too great to allow the agent to act effectively: by time you have realized that a predator is about to attack, it may be too late to escape. This is one of the primary reasons a cognitive system must anticipate future events: so that it can prepare the actions it may need to take in advance of actually sensing that these actions are needed.

In addition to sensory latencies, there are also limitations imposed by the environment and the cognitive system's body. To perform an action, and specifically to accomplish the goal associated with an action, you need to have the relevant part of your body in a certain place at a certain time. It takes time to move, so, again, you need to be able to predict what might happen and prepare to act. For example, if you have to catch an object, you need to start moving your hand before the object arrives and

[14] The concepts of self-maintenance and recursive self-maintenance in self-organizing autonomous system was introduced by Mark Bickhard [13]. We will discuss them in more detail in Chapter 2. The key idea is that self-maintenant systems make active contributions to their own persistence but do not contribute to the maintenance of the conditions for persistence. On the other hand, recursive self-maintenant systems do contribute actively to the conditions for persistence.

[15] When an influence on a system isn't directly controlling it but nonetheless has some impact on the behaviour of the system, we refer to it as a *perturbation*.

[16] The idea of cognition being concerned with *effective action*, i.e. action that helps preserve the system's autonomy, is due primarily to Francisco Varela and Humberto Maturana [14]. These two scientists have had a major impact on the world of cognitive science through their work on biological autonomy and the organizational principles which underpin autonomous systems. Together, they provided the foundations for a new approach to cognitive science called *Enaction*. We will discuss enaction and enactive systems is more detail in Chapter 2.

sometimes even before it has been thrown. Also, the world in which the system is embedded is constantly changing and is outside the control of the system. Consequently, the sensory data which is available to the cognitive system may not only be late in arriving but critical information may also be missing. Filling in these gaps is another of the primary functions of a cognitive system. Paradoxically, it is also often the case that there is too much information for the system to deal with and it has to ignore some of it.[17]

Now, while these capabilities derive directly from the biological autonomy-preserving view of cognition, it should be fairly clear that they would also be of great use to artificial cognitive systems, whether they are autonomous or not. However, before moving on to the next section which elaborates a little more on the relationship between biological and artificial cognitive systems, it is worth noting that some people consider that cognition should involve even more than what we have discussed so far. For example, an artificial cognitive system might also be able to explain what it is doing and why it is doing it.[18] This would enable the system to identify potential problems which could appear when carrying out a task and to know when it needed new information in order to complete it. Taking this to the next level, a cognitive system would be able to view a problem or situation in several different ways and to look at alternative ways of tackling it. In a sense, this is similar to the attribute we discussed above about cognition involving an ability to anticipate the need for actions and their outcomes. The difference in this case is that the cognitive system is considering not just one but *many* possible sets of needs and outcomes. There is also a case to be made that cognition should involve a sense of self-reflection:[19] an ability on the part of the system to think about itself and its own thoughts. We see here cognition straying into the domain of consciousness. We won't say anything more in this book on that subject apart from remarking that computational modelling of consciousness is an active area of research in which the study of cognition plays an important part.

[17] The problem of ignoring information is related to two problems in cogitive science: the *Frame Problem* and *Attention*. We will take up these issues again later in the book.

[18] The ability not simply to act but to explain the reasons for an action was proposed by Ron Brachman in an article entitled "Systems that know what they're doing" [15].

[19] Self-reflection, often referred to as meta-cognition, is emphasized by some people, e.g. Aaron Sloman [16] and Ron Sun [17], as an important aspect of advanced cognition.

1.4   Levels of Abstraction in Modelling Cognitive Systems

All systems can be viewed at different levels of abstraction, successively removing specific details at higher levels and keeping just the general essence of what is important for a useful model of the system. For example, if we wanted to model a physical structure, such as a suspension bridge, we could do so by specifying each component of the bridge — the concrete foundations, the suspension cables, the cable anchors, the road surface, and the traffic that uses it — and the way they all fit together and influence one another. This approach models the problem at a very low level of abstraction, dealing directly with the materials from which the bridge will be built, and we would really only know after we built it whether or not the bridge will stay up. Alternatively, we could describe the forces at work in each member of the structure and analyze them to find out if they are strong enough to bear the required loads with an acceptable level of movement, typically as a function of different patterns of traffic flow, wind conditions, and tidal forces. This approach models the problem at a high level of abstraction and allows the architect to established whether or not his or her design is viable before it is constructed. For this type of physical system, the idea is usually to use an abstract model to validate the design and then realize it as a physical system. However, deciding on the best level of abstraction is not always straightforward. Other types of system — biological ones for example — don't yield easily to this top-down approach. When it comes to modelling cognitive systems, it will come as no surprise that there is some disagreement in the scientific community about what level of abstraction one should use and how they should relate to one another. We consider here two contrasting approaches to illustrate their differences and their relative merits in the context of modelling and designing artificial cognitive systems.

   As part of his influential work on modelling the human visual system, David Marr[20] advocated a three-level hierarchy of abstraction;[21] see Figure 1.6. At the top level, there is the computational theory. Below this, there is the level of representation and algorithm. At the bottom, there is the hardware implementation.

[20] David Marr was a pioneer in the field of computer vision. He started out as a neuroscientist but shifted to computational modelling to try to establish a deeper understanding of the human visual system. His seminal book *Vision* [18] was published posthumously in 1982.

[21] Marr's three-level hierarchy is sometimes known as the *Levels of Understanding* framework.

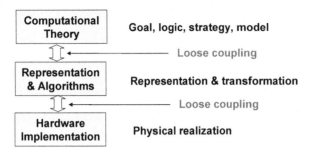

At the level of the computational theory, you need to answer questions such as "what is the goal of the computation, why is it appropriate, and what is the logic of the strategy by which it is carried out?" At the level of representation and algorithm, the questions are different: "how can this computational theory be applied? In particular, what is the representation for the input and output, and what is the algorithm for the transformation?" Finally, the question at the level of hardware implementation is "how can the representation and algorithm be physically realized?" In other words, how can we build the physical system? Marr emphasized that these three levels are only loosely coupled: you can — and, according to Marr, you should — think about one level without necessarily paying any attention to those below it. Thus, you begin modelling at the computational level, ideally described in some mathematical formalism, moving on to representations and algorithms once the model is complete, and finally you can decide how to implement these representations and algorithms to realize the working system. Marr's point is that, although the algorithm and representation levels are more accessible, it is the computational or theoretical level that is critically important from an information processing perspective. In essence, he states that the problem can and should first be modelled at the abstract level of the computational theory without strong reference to the lower and less abstract levels.[22] Since many people believe that cognitive systems — both biological and artificial — are effectively information processors, Marr's hierarchy of abstraction is very useful.

Marr illustrated his argument succinctly by comparing the

Figure 1.6: The three levels at which a system should be understood and modelled: the computational theory that formalizes the problem, the representational and algorithmic level that addresses the implementation of the theory, and the hardware level that physically realizes the system (after David Marr [18]). The computational theory is primary and the system should be understood and modelled first at this level of abstraction, although the representational and algorithmic level is often more intuitively accessible.

[22] Tomaso Poggio recently proposed a revision of Marr's three-level hierarchy in which he advocates greater emphasis on the connections between the levels and an extension of the range of levels, adding *Learning and Development* on top of the computational theory level (specifically hierarchical learning), and *Evolution* on top of that [19]. Tomaso Poggio co-authored the original paper [20] on which David Marr based his more famous treatment in his 1982 book *Vision* [18].

problem of understanding vision (Marr's own goal) to the problem of understanding the mechanics of flight.

> "Trying to understand perception by studying only neurons is like trying to understand bird flight by studying only feathers: it just cannot be done. In order to understand bird flight, we have to understand aerodynamics; only then do the structure of feathers and the different shapes of birds' wings make sense"

Objects with different cross-sectional profiles give rise to different pressure patterns on the object when they move through a fluid such as air (or when a fluid flows around an object). If you choose the right cross-section then there is more pressure on the bottom than on the top, resulting in a lifting force that counters the force of gravity and allows the object to fly. It isn't until you know this that you can begin to understand the problem in a way that will yield a solution for your specific needs.

Of course, you eventually have to decide how to realize a computational model but this comes later. The point he was making is that you should decouple the different levels of abstraction and begin your analysis at the highest level, avoiding consideration of implementation issues until the computational or theoretical model is complete. When it is, it can then subsequently drive the decisions that need to be taken at the lower level when realizing the physical system.

Marr's dissociation of the different levels of abstraction is significant because it provides an elegant way to build a complex system by addressing it in sequential stages of decreasing abstraction. It is a very general approach and can be applied successfully to modelling, designing, and building many different systems that depend on the ability to process information. It also echoes the assumptions made by proponents of a particular paradigm of cognition — *cognitivism* — which we will meet in the next chapter.[23]

Not everyone agrees with Marr's approach, mainly because they think that the physical implementation has a direct role to play in understanding the computational theory. This is particularly so in the emergent paradigm of embodied cognition which we will meet in the next chapter, the embodiment reflecting the physical implementation. Scott Kelso,[24] makes a case for a com-

[23] The cognitivist approach to cognition proposes an abstract model of cognition which doesn't require you to consider the final realization. In other words, cognitivist models can be applied to any platform that supports the required computations and this platform could be a computer or a brain. See Chapter 2, Section 2.1, for more details.

[24] Over the last 25 years, Scott Kelso, the founder of the Center for Complex Systems and Brain Sciences at Florida Atlantic University, has developed a theory of *Coordination Dynamics*. This theory, grounded in the concepts of self-organization and the tools of coupled nonlinear dynamics, incorporates essential aspects of cognitive function, including anticipation, intention, attention, multimodal integration, and learning. His book, *Dynamic Patterns – The Self-Organization of Brain and Behaviour* [21], has influenced research in cognitive science world-wide.

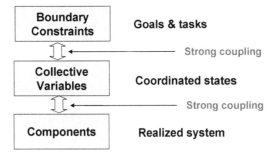

Figure 1.7: Another three
levels at which a system
should be modelled: a
boundary constraint level
that determines the task or
goal, a collective variable
level that characterizes
coordinated states, and a
component level which
forms the realized system
(after Scott Kelso [21]).
All three levels are equally
important and should be
considered together.

pletely different way of modelling systems, especially non-linear
dynamical types of systems that he believes may provide the true
basis for cognition and brain dynamics. He argues that these
types of system should be modelled at three distinct levels of abstraction, but at the same time. These three levels are a boundary
constraint level, a collective variables level, and a components
level. The boundary constraint level determines the goals of
the system. The collective variable[25] level characterizes the behaviour of the system. The component level forms the realized
physical system. Kelso's point is that the specification of these
three levels of model abstraction are tightly coupled and mutually dependent. For example, the environmental context of the
system often determines what behaviours are feasible and useful. At the same time, the properties of the physical system may
simplify the necessary behaviour. Paraphrasing Rolf Pfeifer,[26]
"morphology matters": the properties of the physical shape or
the forced needed for required movements may actually simplify
the computational problem. In other words, the realization of
the system and its particular shape or morphology cannot be
ignored and should not be abstracted away when modelling the
system. This idea that you cannot model the system in isolation
from either the system's environmental context or the system's
ultimate physical realization is linked directly to the relationship
between brain, body, and environment. We will meet it again
later in the book when we discuss enaction in Chapter 2 and
when we consider the issue of embodiment in Chapter 5.

The mutual dependence of system realization and system

[25] Collective variables, also
referred to as order parameters, are so called because
they are responsible for the
system's overall collective
behaviour. In dynamical
systems theory, collective
variables are a small subset of the system's many
degrees of freedom but
they govern the transitions
between the states that the
system can exhibit and
hence its global behaviour.

[26] Rolf Pfeifer, University
of Zurich, has long been
a champion of the tight
relationship between a
system's embodiment and
its cognitive behaviour, a
relationship set out in his
book *How the body shapes the
way we think: A new view of
intelligence* [22], co-authored
by Josh Bongard.

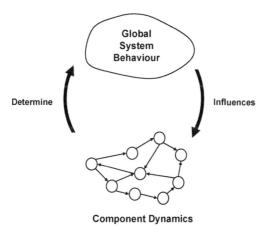

Figure 1.8: Circular causality — sometimes referred to as continuous reciprocal causation or recursive self-maintenance — refers to the situation where global system behaviour somehow influences the local behaviour of the system components and yet it is the local interaction between the components that determines the global behaviour. This phenomenon appears to be one of the pivotal mechanisms in autonomous cognitive systems.

modelling presents us with a difficulty, however. If we look carefully, we see a circularity, with everything depending on something else. It's not easy to see how you break into the modelling circle. This is one of the attractions of Marr's approach: there is a clear place to get started. This circularity crops up repeatedly in cognition and it does so in many forms. All we will say for the moment is that circular causality[27] — where global system behaviour somehow influences the local behaviour of the system components and yet it is the local interaction between the components that determines the global behaviour; see Figure 1.8 — appears to be one of the key mechanisms of cognition. We will return again to this point later in the book. For the moment, we'll simply remark that the two constrasting approaches to system modelling mirror two opposing paradigms of cognitive science. It is to these that we now turn in Chapter 2 to study the foundations that underpin our understanding of natural and artificial cognitive systems.

[27] Scott Kelso uses the term "circular causality" to describe the situation in dynamical systems where the cooperation of the individual parts of the system determine the global system behaviour which, in turn, governs the behaviour of these individual parts [21]. This is related to Andy Clark's concept of continuous reciprocal causation (CRC) [23] which "occurs when some system S is both continuously affecting and simultaneously being affected by, activity in some other system O" [24]. These ideas are also echoed in Mark Bickhard's concept of recursive self-maintenance [13]. We will say more about these matters in Chapter 4.

2

Paradigms of Cognitive Science

In Chapter 1, we were confronted with the tricky and unexpected problem of how to define cognition. We made some progress by identifying the main characteristics of a cognitive system — perception, learning, anticipation, action, adaptation, autonomy — and we introduced four aspects that must be borne in mind when modelling a cognitive system: (a) biological inspiration *vs.* computational theory, (b) the level of abstraction of the model, (c) the mutual dependence of brain, body, and environment, and (d) the ultimate-proximate distinction between what cognition is for and how it is achieved. However, we also remarked on the fact that there is more than one tradition of cognitive science so that any definition of cognition will be heavily coloured by the background against which the definition is set. In this chapter, we will take a detailed look at these various traditions. Our goal is to tease out their differences and get a good grasp of what each one stands for. Initially, it will seem that these traditions are polar opposites and, as we will see, they do differ in many ways. However, as we get to the end of the chapter, we will also recognize a certain resonance between them. This shouldn't surprise us: after all, each tradition occupies its own particular region of the space spanned by the ultimate and proximate dimensions which we discussed in Chapter 1 and it is almost inevitable that there will be some overlap, especially if that tradition is concerned with a general understanding of cognition.

Before we begin, it's important to appreciate that cognitive sci-

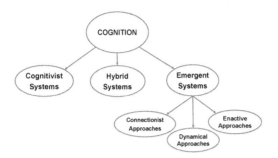

Figure 2.1: The cognitivist, emergent, and hybrid paradigms of cognition.

ence is a general umbrella term that embraces several disciplines, including neuroscience, cognitive psychology, linguistics, epistemology, and artificial intelligence, among others. Its primary goal is essentially to understand and explain the underlying processes of cognition: typically human cognition and ideally in a way that will yield a model of cognition that can then be replicated in artificial agents.

To a large extent, cognitive science has its origins in cybernetics which in the early 1940s to 1950s made the first attempts to formalize what had up to that point been purely psychological and philosophical treatments of cognition. Cybernetics was defined by Norbert Wiener as "the science of control and communication in the animal and the machine."[1] The intention of the early cyberneticians was to understand the mechanisms of cognition and to create a science of mind, based primarily on logic. Two examples of the application of cybernetics to cognition include the seminal paper by Warren S. McCulloch and Walter Pitts "A logical calculus immanent in nervous activity"[2] and W. Ross Ashby's book *Design for a Brain*.[3]

The first attempts in cybernetics to uncover the mechanisms of cognitive behaviour were subsequently taken up and developed into an approach referred to as *cognitivism*. This approach built on the logical foundations laid by the early cyberneticians and exploited the newly-invented computer as a literal metaphor for cognitive function and operation, using symbolic information processing as its core model of cognition. The cognitivist tradition continued to be the dominant approach over the next

[1] The word *cybernetics* has its roots in the Greek word κυβερνήτης or kybernētēs, meaning steersman. It was defined in Norbert Wiener's book *Cybernetics* [25], first published in 1948, as "the science of control and communication" (this was the sub-title of the book). W. Ross Ashby remarks in his book *An Introduction to Cybernetics* [26], published in 1956, that cybernetics is essentially "the art of steersmanship" and as such its themes are co-ordination, regulation, and control.

[2] As well as being a seminal work in cybernetics, the 1943 paper by Warren S. McCulloch and Walter Pitts, "A logical calculus immanent in nervous activity" [27], is also regarded as the foundation for artificial neural networks and connectionism [28].

[3] Complementing his influential book *Design for a Brain* [29, 30, 31], W. Ross Ashby's *Introduction to Cybernetics'* [26] is also a classic text.

30 or so years and, indeed, it was so pervasive and became so deeply embedded in our mind-set that it still holds sway today to a considerable extent.

Paradoxically, the early cybernetics period also paved the way for a completely different approach to cognitive science — the emergent systems approach — which recognized the importance of self-organization in cognitive processes. Initially, emergent systems developed almost under the radar — it was difficult to challenge the appeal of exciting new computer technology and the computational model of cognition — but it progressed nonetheless in parallel with the cognitivist tradition over the next fifty years and more, growing to embrace connectionism, dynamical systems theory, and enaction, all of which we will discuss in more detail later in the chapter.

In recent years, a third class — hybrid systems — has become popular, and understandably so since, as the name suggests, it attempts to combine the best from each of the cognitivist and emergent paradigms; see Figure 2.1.

In the sections that follow, we will take a closer look at all three traditions of cognitive science — cognitivist, emergent, and hybrid — to draw out the key assumptions on which they build their respective theories and to compare and contrast them on the basis of several fundamental characteristics that reflect different points in the ultimate-proximate space. Before we proceed to do this, it is worth noting that, although we have referred so far to the different *traditions* of cognitive science, the title of the chapter refers to the different *paradigms* of cognitive science. Is there any great significance to this? Well, in fact, there is. As we will see in what follows, and not withstanding the resonance that we mentioned above, the two traditions do make some fundamentally different assertions about the nature of cognition (i.e. its ultimate purpose) and its processes (i.e. its proximate mechanisms). In fact, these differences are so strong as to render the two approaches intrinsically incompatible and, hence, position them as two completely different paradigms. It isn't hard to see that this incompatibility is going to cause problems for hybrid approaches, but we'll get to that in due course. For the present, let's proceed with our discussion of the cognitivist and emergent traditions of cognitive science, seeking out the issues on which they agree, but recognizing too those on which they do not, both in principle and in practice.[4]

2.1   The Cognitivist Paradigm of Cognitive Science

2.1.1   An Overview of Cognitivism

As we have seen, the initial attempt in cybernetics to create a science of cognition was followed by the development of an approach referred to as cognitivism. The birth of the cognitivist paradigm, and its sister discipline of Artificial Intelligence (AI), dates from a conference held at Dartmouth College, New Hampshire, in July and August 1956. It was attended by people such as John McCarthy, Marvin Minsky, Allen Newell, Herbert Simon, and Claude Shannon, all of whom had a very significant influence on the development of AI over the next half-century. The essential position of cognitivism is that cognition is achieved by computations performed on internal symbolic knowledge representations in a process whereby information about the world is taken in through the senses, filtered by perceptual processes to generate descriptions that abstract away irrelevant data, represented in symbolic form, and reasoned about to plan and execute mental and physical actions. The approach has also been labelled by many as the information processing or symbol manipulation approach to cognition.

For cognitivist systems, cognition is representational in a particular sense: it entails — requires — the manipulation of explicit symbols: localized abstract encapsulations of information that denote the state of the world external to the cognitive agent. The term 'denote' has particular significance here because it asserts an identity between the symbol used by the cognitive agent and the thing that it denotes. It is as if there is a one-to-one correspondence between the symbol in the agent's cognitive system and the state of the world to which it refers. For example, the clothes in a laundry basket can be represented by a set of symbols, often organized in a hierarchical manner, describing the identity of each item and its various characteristics: whether it is

[4] For an accessible summary of the different paradigms of cognition, refer to a paper entitled "Whence Perceptual Meaning? A Cartography of Current Ideas" [32] It was written by Francisco Varela, one of the founders of a branch of cognitive science called Enaction, and it is particularly instructive because, as well as contrasting the various views of cognition, it also traces them to their origins. This historical context helps highlight the different assumptions that underpin each approach and it shows how they have evolved over the past sixty years or so. Andy Clark's book *Mindware – An Introduction to the Philosophy of Cognitive Science* [33] also provides a useful introduction to the philosophical and scientific differences between the different paradigms of cognition.

heavily soiled or not, its colour, its recommended wash cycle and water temperature. Similarly, the washing machine can be represented by another symbol or set of symbols. These symbols can represent objects and events but they can also represent actions: things that can happen in the world. For example, symbols that represent sorting the clothes into bundles, one bundle for each different wash cycle, putting them into the washing machine, selecting the required wash cycle, and starting the wash. It is a very clear, neat, and convenient way to describe the state of the world in which the cognitive agent finds itself.

Having this information about the world represented by such an explicit abstract symbolic knowledge is very useful for two reasons. First, it means that you can easily combine this knowledge by associating symbolic information about things and symbolic information about actions that can be performed on them and with them. These associations effectively form rules that describe the possible behaviours of the world and, similarly, the behaviours of the cognitive agent. This leads to the second reason why such a symbolic representational view of cognition is useful: the cognitive agent can then reason effectively about this knowledge to reach conclusions, make decisions, and execute actions. In other words, the agent can make inferences about the world around it and how it should behave in order to do something useful, i.e. to perform some task and achieve some goal. For example, if a particular item of clothing is heavily soiled but it is a delicate fabric, the agent can select a cool wash cycle with a pre-soak, rather than putting it into a hot water cycle (which will certainly clean the item of clothing but will probably also cause it to shrink and fade). Of course, the agent needs to know all this if it is to make the right decisions, but this doesn't present an insurmountable difficulty as long as someone or something can provide the requisite knowledge in the right form. In fact, it turns out to be relatively straightforward because of the denotational nature of the knowledge: other cognitive agents have the same representational framework and they can share this domain knowledge directly[5] with the cognitive robot doing the laundry. This is the power of the cognitivist perspective on cognition and knowledge.

[5] The idea of cognitive robots sharing knowledge is already a reality. For example, as a result of the RoboEarth initiative, robots are now able to share information on the internet: see the article by Markus Waibel and his colleagues "RoboEarth: A World-Wide Web for Robots" [34]. For more details, see Chapter 8, Section 8.6.1.

A particular feature of this shared symbolic knowledge —
rules that describe the domain in which the cognitive agent is
operating — is that it is even more abstract than the symbolic
knowledge the agent has about its current environment: this
domain knowledge describes things *in general*, in a way that isn't
specific to the particular object the agent has in front of it or the
actions it is currently peforming. For example, the knowledge
that delicate coloured fabrics will fade and shrink in very hot
water isn't specific to the bundle of laundry that the agent is
sorting but it does apply to it nonetheless and, more to the point,
it can be used to decide how to wash this particular shirt.

Now, let's consider for a moment the issue of where the
agent's knowledge comes from. In most cognitivist approaches
concerned with creating artificial cognitive systems, the symbolic
knowledge is the descriptive product of a human designer. The
descriptive aspect is important: the knowledge in question is
effectively a description of how the designer — a third-party observer — sees or comprehends the cognitive agent and the world
around it. So, why is this a problem? Well, it's not a problem if
every agent's description is identical or, at the very least, compatible: if every agent sees and experiences the world the same
way and, more to the point, generates a compatible symbolic
representation of it. If this is the case — and it will be the case if
the assertion which cognitivism makes that an agent's symbolic
representation denotes the objects and events in the world is
true — then the consequence is very significant because it means
that these symbolic representations can be directly accessed,
understood, and shared by the cognitive agent (including other
people). Furthermore, it means that domain knowledge can be
embedded directly in to, and extracted from, an artificial cognitive agent. This direct transferrability of knowledge between
agents is one of cognitivism's key characteristics. Clearly, this
makes cognitivism very powerful and extremely appealing, and
the denotational attribute of symbolic knowledge is one of its
cornerstones.

You may have noticed above a degree of uncertainty in the
conditional way we expressed the compatibility of descriptive knowledge and its denotational quality ("if the assertion

... that an agent's symbolic representation denotes the objects and events in the world is true"). Cognitivism asserts that this is indeed the case but, as we will see in Section 2.2, the emergent paradigm of cognitive science takes strong issue with this position. Since this denotational aspect of knowledge and knowledge sharing (in effect, cognitivist epistemology) is so important, it will come as no surprise that there are some far-reaching implications. One of them concerns the manner in which cognitive computations are carried out and, specifically, the issue of whether or not the platform that supports the required symbolic computation is of any consequence. In fact, in the cognitivist paradigm, it isn't of any consequence: any physical platform that supports the performance of the required symbolic computation will suffice. In other words, a given cognitive system (technically, for a given cognitive architecture; but we'll wait until Chapter 3 to discuss this distinction) and its component knowledge (the content of the cognitive architecture) can exploit any machine that is capable of carrying out the required symbol manipulation. This could be a human brain or a digital computer. The principled separation of computational operation from the physical platform that supports these computations is known as *computational functionalism*.[6] Cognitivist cognitive systems are computationally functionalist systems.

Although the relationship between computational functionalism and the universal capability of cognitivist symbolic knowledge is intuitively clear — if every cognitive agent has the same world view and a compatible representational framework, then the physical support for the symbolic knowledge and associated computation is of secondary importance — they both have their roots in classical artificial intelligence, a topic to which we now turn. We will return to the cognitivist perspective on knowledge and representation in Chapter 8.

2.1.2   *Cognitivism and Artificial Intelligence*

As we mentioned at the outset, both cognitivist cognitive science and artificial intelligence share a common beginning and they developed together, building a strong symbiotic relationship over

[6] The principled decoupling of the computational model of cognition from its instantiation as a physical system is referred to as *computational functionalism*. It has its roots in, for example, Allen Newell's and Herbert Simon's seminal paper "Computer Science as Empirical Enquiry: Symbols and Search" [35]. The chief point of computational functionalism is that the physical realization of the computational model is inconsequential to the model: any physical platform that supports the performance of the required symbolic computations will suffice, be it computer or human brain; also see Chapter 5, Section 5.2.

a period of approximately thirty years.[7] Artificial intelligence then diverged somewhat from its roots, shifting its emphasis away from its orginal concern with human and artificial cognition and their shared principles to issues concerned more with practical expediency and purely computational algorithmic techniques such as statistical machine learning. However, the past few years have seen a return to its roots in cognitivist cognitive science, now under the banner of Artificial General Intelligence (to reflect the reassertion of the importance of non-specific approaches built on human-level cognitive foundations).[8] Since there is such a strong bond between cognitivism and classical artificial intelligence, it is worth spending some time discussing this relationship.

In particular, because it has been extraordinarily influential in shaping how we think about intelligence, natural as well as computational, we will discussAllen Newell's and Herbert Simon's "Physical Symbol System" approach to artificial intelligence.[9] As is often the case with seminal writing, the commentaries and interpretations of the original work frequently present it in a somewhat distorted fashion and some of the more subtle and deeper insights get lost. Despite our brief treatment, we will try to avoid this here.

Allen Newell and Herbert Simon, in their 1975 ACM Turing Award Lecture "Computer Science as Empirical Enquiry: Symbol and Search," present two hypotheses:

1. The *Physical Symbol System Hypothesis:* A physical symbol system has the necessary and sufficient means for general intelligent action.

2. The *Heuristic Search Hypothesis:* The solutions to problems are represented as symbol structures. A physical-symbol system exercises its intelligence in problem-solving by search, that is, by generating and progressively modifying symbol structures until it produces a solution structure.

The first hypothesis implies that any system that exhibits general intelligence is a physical symbol system and, furthermore, any physical symbol system of sufficient size can be configured somehow ("organized further," in the words of Newell and

[7] Some observers view AI less as a discipline that co-developed along with cognitivist cognitive science and more as the direct descendent of cognitivism. Consider the following statement by Walter Freeman and Rafael Núñez: "... the positivist and reductionist study of the mind gained an extraordinary popularity through a relatively recent doctrine called *Cognitivism*, a view that shaped the creation of a new field — *Cognitive Science* — and its most hard core offspring: Artificial Intelligence" (emphasis in the original) [36].

[8] The renewal of the cognitivist goals of classical artificial intelligence to understand and model human-level intelligence is typified by the topics addressed by the cognitive systems track of the AAAI conference [37], and the emergence of the discipline of *artificial general intelligence* promoted by, among others, the Artificial General Intelligence Society [38] and the Artificial General Intelligence Research Institute [39].

[9] Allen Newell and Herbert Simon were the recipients of the 1975 ACM Turing Award. Their Turing Award lecture "Computer Science as Empirical Enquiry: Symbol and Search" [35] proved to be extremely influential in the development of artificial intelligence and cognitivist cognitive science.

Simon) to exhibit general intelligence. This is a very strong assertion. It says two things: (a) that it is necessary for a system to
be a physical symbol system if the system is to display general
intelligence (or cognition), and (b) that being a physical symbol
system of adequate size is sufficient to be an intelligent system
— you don't need anything else.

The second hypothesis amounts to an assertion that symbol systems solve problems by heuristic search, *i.e.* "successive
generation of potential solution structures" in an effective and
efficient manner: "The task of intelligence, then, is to avert the
ever-present threat of the exponential explosion of search." This
hypothesis is sometimes caricatured by the statement that "All
AI is search" but this is to unfairly misrepresent the essence of
the second hypothesis. The point is that a physical symbol system must indeed search for solutions to the problem but it is
intelligent because its search strategy is effective and efficient:
it doesn't fall back into blind exhaustive search strategies that
would have no hope of finding in a reasonable amount of time a
solution to the kinds of problems that AI is interested in. Why?
Because these are exactly the problems that defy simple exhaustive search techniques by virtue of the fact that the computational complexity of these brute-force solutions — the amount
of time needed to solve them — increases exponentially with
the size of the problem.[10] It is in this sense that the purpose of
intelligence is to deal effectively with the danger of exponentially
large search spaces.

A physical symbol system is essentially a machine that produces over time an evolving collection of symbol structures.[11] A
symbol is a physical pattern and it can occur as a component of
another type of entity called an expression or symbol structure:
in other words, expressions or symbol structures are arrange-
ments of symbols. As well as the symbol structures, the system
also comprises processes that operate on expressions to produce
other expressions: to process, create, modify, reproduce, and de-
stroy them. An expression can *designate* an object and thereby
the system can either affect the object itself or behave in ways
that depend on the object. Alternatively, if the expression des-
ignates a process, then the system *interprets* the expression by

[10] Formally, we say that
exponential complexity is
of the order k^n, where k is
a constant and n is the size
of the problem. By contrast,
we also have polynomial
complexity: n^2, n^3, or n^k.
The difference between these
two classes is immense.
Problems with exponential
complexity solutions scale
very badly and, for any
reasonably-sized problem,
are usually intractable, i.e.
they can take days or years
to solve, unless some clever
— intelligent — solution
strategy is used.

[11] For a succinct overview of
symbol systems, see Stevan
Harnad's seminal article
"The Symbol Grounding
Problem" [40].

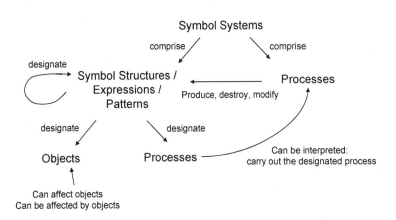

Figure 2.2: The essence of a physical symbol system [35].

carrying out the process (see Figure 2.2). In the words of Newell and Simon,

> "Symbol systems are collections of patterns and processes, the latter being capable of producing, destroying, and modifying the former. The most important properties of patterns is that they can designate objects, processes, or other patterns, and that when they designate processes, they can be interpreted. Interpretation means carrying out the designated process. The two most significant classes of symbol systems with which we are acquainted are human beings and computers."

There is an important if subtle point here. This explanation of a symbol system is much more general and powerful than the usual portrayal of symbol-manipulation systems in which symbols designate only objects, and in which case we have a system of processes that produces, destroys, and modifies symbols, and no more. Newell's and Simon's original view is considerably more sophisticated. There are two recursive aspects to it: processes can produce processes, and patterns can designate patterns (which, in turn, can be processes). These two recursive loops are closely linked. Not only can the system build ever more abstract representations and reason about those representation, but *it can modify itself* as a function of its processing and its symbolic representations. The essential point is that in Newell's and Simon's original vision, physical symbol systems are in principle capable of development. This point is often lost in contemporary discussions of cognitivist AI systems. On the other hand, as we will see later, emergent approaches focus squarely on the need for development. This is one of the resonances between the cognitivist and emergent paradigms we mentioned, although one that isn't often picked up on because the developmental capacity that is intrinsic in principle to cognitivist system, by virtue of the physical symbol systems hypothesis, often doesn't get the recognition it deserves.

In order to be realized as a practical agent, symbol systems have to be executed on some computational platform. However, the behaviour of these realized systems depend on the details of the symbol system, its symbols, operations, and interpretations, and *not* on the particular form of the realization. This is something we have already met: the functionalist nature of cognitivist cognitive systems. The computational platform that supports the physical symbol system or the cognitive model is arbitrary to the extent that it doesn't play any part in the model itself. It may well influence how fast the processes run and the length of time it takes to produce a solution, but this is only a matter of timing and not outcome, which will be the same in every case.

Thus, the physical symbol system hypothesis asserts that a physical symbol system has the necessary and sufficient means for general intelligence. From what we have just said about symbol systems, it follows that intelligent systems, either natural or artificial ones, are effectively equivalent because the instantiation is actually inconsequential, at least in principle. It is evident that, to a very great extent, cognitivist systems and physical symbol systems are effectively identical with one another. Both share the same assumptions, and view cognition or intelligence in exactly the same way.

Shortly after Newell and Simon published their influential paper, Allen Newell defined intelligence as the degree to which a system approximates the ideal of a knowledge-level system.[12] This is a system which can bring to bear *all* its knowledge onto *every* problem it attempts to solve (or, equivalently, every goal it attempts to achieve). Perfect intelligence implies complete utilization of knowledge. It brings this knowledge to bear ac-

[12] In addition to his seminal 1975 Turing Award Lecture, Allen Newell made several subsequent landmark contributions to the establishment of practical cognitivist system, beginning perhaps in 1982 with his introduction of the concept of a knowledge-level system, the Maximum Rationality Hypothesis, and the principle of rationality [41], in the mid-1980s with the development of the Soar cognitive architecture for general intelligence (along with John Laird and Paul Rosenbloom) [42], and in 1990 the idea of a Unified Theory of Cognition [43].

cording to the Maximum Rationality Hypothesis expressed as
the *principle of rationality* which was proposed by Allen Newell
in 1982 as follows: "If an agent has knowledge that one of its
actions will lead to one of its goals, then the agent will select that
action." John Anderson later offered a slightly different perspective, referred to as *rational analysis*, in which the cognitive system
optimizes the adaptation of the behaviour of the organism. Note
that Anderson's principle considers optimality to be necessary
for rationality, something that Newell's principle does not.[13]
In essence, the principle of rationality formalizes the intuitive
idea that an intelligent agent will never ignore something if it
knows it will help achieve its goal and will always use as much
of its knowledge as it can to guide its behaviour and successfully
complete whatever task it is engaged in.

[13] For a good comparison of Newell's principle of ratio-nality [41] and Anderson's rational analysis [44], refer to the University of Michigan's Survey of Cognitive and Agent Architectures [45].

As we might expect, the knowledge in such an artificial intelligence system, i.e. in a knowledge-level system, is represented
by symbols. Symbols are abstract entities that may be instantiated and manipulated as *tokens*. Developing his physical symbol
systems hypothesis, Newell characterizes a symbol system as
follows.[14] It has:

[14] Newell's characterization of a symbol system can be found on a website at the University of Michigan dedicated to cognitive and agent architectures [45] .

- *Memory* to contain the symbolic information;

- *Symbols* to provide a pattern to match or index other symbolic
  information;

- *Operations* to manipulate symbols;

- *Interpretations* to allow symbols to specify operations;

- *Capacities* for *composability*, so that the operators may produce
  any symbol structure; for *interpretability*, so that the symbol
  structures are able to encode any meaningful arrangement
  of operations; and sufficient *memory* to facilitate both of the
  above.

Newell suggests a progression of four bands of operation,
depending on the timescale over which processing takes place,
ranging from biological, cognitive, rational, to social. The typical
execution time in the biological band is 10^{-4} to 10^{-2} seconds,
the cognitive 10^{-1} to 10^{1} seconds, the rational 10^{2} to 10^{4} seconds,

and the social 10^5 to 10^7 seconds. The biological band corresponds to the neurophysiological make-up of the system. Newell identifies three layers in this band: the organelle, the neuron, and the neural circuit. Connectionist systems and artifical neural networks are often focussed exclusively on this band.

The cognitive band corresponds to the symbol level and its physical instantiation as a concrete architecture. The idea of a cognitive architecture is one of the most important topics in the study of cognitive systems (both biological and artificial) and artificial intelligence, and cognitive architectures play a key role in cognitivist cognitive science in particular. We devote all of the next chapter to this topic.

Newell identifies three layers in the cognitive band. First, there are deliberate acts that take a very short amount of time, typically 100ms. For example, reaching to grasp an object. Second, there are 'composed operations' which comprise sequences of deliberate acts. For example, reaching for an article of clothing, grasping it, picking it up, and putting it in a washing machine. These composed operations take on the order of a second. Third, there are complete actions that take up to ten seconds. For example, finding the washing powder tablets, opening the washing powder tray in the washing machine, inserting the tablet, and adding the fabric conditioner.

The rational band is concerned with actions that are typically characterized by tasks and require some reasoning. For example, doing the laundry. The social band extends activity to behaviours that occupy hours, days, or weeks, often involving interaction with other agents.

All knowledge is represented (symbolically) at the cognitive symbol level. All knowledge-level systems contain a symbol system. As we have already seen, this is the strong interpretation of the physical symbol system hypothesis: not only is a physical symbol system *sufficient* for general intelligence, it is also *necessary* for intelligence.

This section has summarized very briefly the close relationship between classical AI and cognitivism and, by extension, the new incarnation of classical AI in Artificial General Intelligence (AGI). It is impossible to overestimate the influence that AI has had on cognitivist cognitive science and, to a slightly lesser extent, that of cognitivism on AI. This is not surprising when you consider that they were both born as disciplines at the same time by more or less the same people and that their goals were identical: to develop a comprehensive theory — inspired by the computational model of information processing — that moved forward the original agenda of the early cyberneticians to formalize the mechanisms that underpin cognition in animals and machines. AI may well have deviated from that goal in the last 20 or so years to pursue alternative strategies such as statistical machine learning but, as we mentioned above, there is now a large and growing body of people who are championing a return to the original goals of the founders of cognitivism and classical AI which, in the words of Pat Langley at Arizona State University, was to understand and reproduce in computational systems the full range of intelligent behaviour observed in humans.

With this important pillar of cognitive science now firmly established, let's move on to the second pillar that also grew out of the goals and aspirations of the early cyberneticians: emergent systems.

2.2   *The Emergent Paradigm of Cognitive Science*

The view of cognition taken by emergent approaches is very different to that taken by cognitivism. The ultimate goal of an emergent cognitive system is to maintain its own autonomy, and cognition is the process by which it accomplishes this. It does so through a process of continual self-organization whereby the agent interacts with the world around it but only in such a way as not to threaten its autonomy. In fact, the goal of cognition is to make sure that the agent's autonomy is not compromised but is continually enhanced to make its interactions increasingly more robust. In achieving this, the cognitive process determines what is real and meaningful for the system: the system constructs its reality — its world and the meaning of its perceptions and actions — as a result of its operation in that world. Consequently, the system's understanding of its world is inherently specific to the form of the system's embodiment and is dependent on the system's history of interactions, i.e., its experiences.[15] This process of making sense of its environmental interactions is one of the foundations of a branch of cognitive science called *enaction*, about which we will say much more in Section 2.2.3. Cognition is also the means by which the system compensates for the immediate "here-and-now" nature of perception, allowing it to anticipate events that occur over longer timescales and prepare for interaction that may be necessary in the future. Thus, cognition is intrinsically linked with the ability of an agent to act prospectively: to deal with what might be, not just with what is.

Many emergent approaches also adhere to the principle that the primary mode of cognitive learning is through the acquisition of new anticipatory skills rather than knowledge, as is the case in cognitivism.[16] As a result, in contrast to cognitivism, emergent approaches are necessarily embodied and the physical form of the agent's body plays a pivotal role in the cognitive process. Emergent systems wholeheartedly embrace the idea of the interdependence between brain, body, and world that we mentioned in the previous chapter. Because of this, cognition in the emergent paradigm is often referred to as *embodied cognition*.[17] However, while the two terms might be synonymous, they are not equivalent. Embodied cognition focusses on the fact that the body and the brain, together, form the basis of a cognitive system and they do so in the context of a structured environmental niche to which the body is adapted. Emergent systems, as we will see, do so too but often they make even stronger assertions about the nature of cognition.

The emergent paradigm can be sub-divided into three approaches: connectionist systems, dynamical systems, and enactive systems (refer again to Figure 2.1). However, it would be wrong to suggest that these three approaches have nothing to do with one another. On the contrary, there are very important ways in which they overlap. The ultimate-proximate relationship again helps to clarify the distinctions we make between them. Both connectionist and dynamical systems are more concerned with proximate explanations of cognition, i.e. the mechanisms by which cognition is achieved in a system. Typically, connectionist systems correspond to models at a lower level of abstraction, dy-

[15] This mutual specification of the system's reality by the system and its environment is referred to as co-determination [14] and is related to the concept of radical constructivism [46] (see Chapter 8, Section 8.3.4).

[16] Emergent approaches typically proceed on the basis that processes which guide action and improve the capacity to guide action form the root capacity of all intelligent systems [47].

[17] We discuss the issue of *embodied cognition* in detail in Chapter 5.

namical systems to a higher level. Enaction, on the other hand, makes some strong claims about what cognition is for (as does the emergent paradigm in general) and why certain characteristics are important. Enaction does have a lot to say about how the process of cognition is effected, touching on the proximate explanations, but it does so at quite an abstract level. To date, explicit formal mechanistic, computational, or mathematical models of enaction remain goals for future research.[18] In contrast, connectionism and dynamical systems theory provide us with very detailed and well-understood formal techniques, mathematically and computationally, but the challenge of scaling them to a fully-fledged theory of cognition on a par with, say, the cognitivist Unified Theory of Cognition (a concept that was already mentioned above in Section 2.1.2 and that will be discussed more fully in Chapter 3) requires much more time and effort, not to mention some intellectual breakthroughs.

Many working on the area feel that the future of the emergent paradigm may lie in the unification of connectionist, dynamical, and enactive approaches, binding them together in a cohesive joint ultimate-proximate explanation of cognition. Indeed, as we will see shortly, connectionism and dynamical systems theory are best viewed as two complementary views of a common approach, the former dealing with microscopic aspects and the latter with macroscopic. On the other hand, others working in the field prefer the view that a marriage of cognitivist and emergent approaches is the best way forward, as exemplified by the hybrid systems approach about which we will say more in Section 2.3. There are other interesting perspectives too, such as the computational mechanics espoused by James Crutchfield who argues for a synthesis and extension of dynamical and information processing approaches.[19]

Bearing in mind these relationships, let us now proceed to examine the three emergent approaches, one by one, highlighting the areas where they overlap.

[18] *Enaction: Towards a New Paradigm for Cognitive Science* [48], edited by John Stewart, Olivier Gapenne, and Ezequiel Di Paolo, and published by MIT Press in 2010, provides an excellent snapshot of the current state of development of the enactive paradigm.

[19] James Crutchfield agrees with those who advocate a dynamical perspective on cognition, asserting that time is a critical element, and he makes the point that one of the advantages of dynamical systems approaches is that it renders the temporal aspects geometrically in a state space. Structures in this state space both generate and constrain behaviour and the emergence of spatio-temporal patterns. Dynamics, then, are certainly involved in cognition. However, he argues that dynamics *per se* are "not a substitute for information processing and computation in cognitive processes" but neither are the two approaches incompatible [49]. He holds that a synthesis of the two can be developed to provide an approach that does allow dynamical state space structures to support computation. He proposes *computational mechanics* as the way to tackle this synthesis of dynamics and computation.

2.2.1 Connectionist Systems

Connectionist systems rely on parallel processing of non-symbolic distributed activation patterns in networks of relatively simple processing elements. They use statistical properties rather than logical rules to analyze information and produce effective behaviour. In the following, we will summarize the main principles of connectionism, briefly tracing its history and highlighting the main developments that have led us to where we are today. Unfortunately, but inevitably, we will be forced into making use of many technical terms with little or no explanation: to do justice to connectionism and the related field of artificial neural networks would require a substantial textbook in its own right. All we can hope for here is to convey some sense of the essence of the connectionism, its relevance to cognitive science, and the way it differs from cognitivism. References to supplementary material are provided in the sidenote.[20]

The roots of connectionism reach back well before the computational era. Although the first use of connectionism for computer-based models dates from 1982,[21] the term connectionism had been used in psychology as early as 1932.[22] Indeed, connectionist principles are clearly evident in William James' nineteenth century model of associative memory,[23] a model that also anticipated mechanisms such as Hebbian learning, an influential unsupervised neural training process whereby the synaptic strength — the bond between connecting neurons — is increased if both the source and target neurons are active at the same time. The introduction to Donald Hebb's book *The Organization of Behaviour* published in 1949 also contains one of the first usages of the term connectionism.[24]

We have already noted that cognitivism has some of its roots in earlier work in cybernetics and in the seminal work by Warren McCulloch and Walter Pitts in particular.[25] They showed that any statement within propositional logic could be represented by a network of simple processing units, i.e. a connectionist system. They also showed that such nets have, in principle, the computational power of a Universal Turing Machine, the theoretical basis for all computation. Thus, McCulloch and Pitts managed

[20] David Medler's paper "A Brief History of Connectionism" [50] provides an overview of classical and contemporary approaches and a summary of the link between connectionism and cognitive science. For a selection of seminal papers on connectionism, see James Anderson's and Edward Rosenfeld's *Neurocomputing: Foundations of Research* [28] and *Neurocomputing 2: Directions of Research* [51]. Paul Smolensky reviews the field from a mathematical perspective [52, 53, 54, 55]. Michael Arbib's *Handbook of Brain Theory and Neural Networks* provides very accessible summaries of much of the relevant literature [56].

[21] The introduction of the term *connectionist models* in 1982 is usually attributed to Jerome Feldman and Dana Ballard in their paper "Connectionist Models and their Properties" [57].

[22] Edward Thorndike used the term connectionism to refer to an extended form of associationism in 1932 [58, 59].

[23] Anderson's and Rosenfeld's collection of papers [28] opens with Chapter XVI "Association" from William James's 1890 *Psychology, Briefer Course* [60].

[24] The introduction of Donald Hebb's book *The Organization of Behaviour* [61] can be found in Anderson's and Rosenfeld's collection of papers [28].

[25] See Sidenote 2 in this chapter.

the remarkable feat of contributing simultaneously to both the foundation of cognitivism and the foundation of connectionism.

The connectionist approach was advanced significantly in the late 1950s with the introduction of Frank Rosenblatt's perceptron and Oliver Selfridge's Pandemonium model of learning.[26] Rosenblatt showed that any pattern classification problem expressed in binary notation can be solved by a perceptron network, a simple network of elementary computing elements which do little more than sum the strength of suitably-weighted input signals or data streams, compare the result to some fixed threshold value, and, on the basis of the result, they either fire or not, producing a single output which then connected to other computing element in the network.

Although network learning advanced in 1960 with the introduction of the Widrow-Hoff rule (also called the delta rule) for supervised training in the *Adaline* neural model,[27] perceptron networks suffered from a severe problem: no learning algorithm existed to allow the adjustment of the weights of the connections between input units and hidden associative units in networks with more than two layers.

In 1969, Marvin Minsky and Seymour Papert caused something of a stir by showing that these perceptrons can only be trained to solve linearly separable problems and couldn't be trained to solve more general problems.[28] As a result, research on neural networks and connectionism suffered considerably.

With the apparent limitations of perceptions clouding work on network learning, research focussed more on memory and information retrieval and, in particular, on parallel models of associative memory.[29]

During this period, alternative connectionist models were also being developed, such as Stephen Grossberg's Adaptive Resonance Theory (ART)[30] and Teuvo Kohonen's self-organizing maps (SOM), often referred to simply as Kohonen networks.[31] ART addresses real-time supervised and unsupervised category learning, pattern classification, and prediction, while Kohonen networks exploit self-organization for unsupervised learning and can be used as either an auto-associative memory or a pattern classifier.

[26] The perceptron was introduced in 1958 by Frank Rosenblatt [62] while the *Pandemonium* learning model was developed by Oliver Selfridge in 1959 [63]; both are included in Anderson's and Rosenfeld's collection of seminal papers [28].

[27] Adaline — for Adaptive Linear — was introduced by Bernard Widrow and Marcian Hoff in 1960 [64].

[28] *Perceptrons: An Introduction to Computational Geometry* by Marvin Minsky and Seymour Papert [65] was published in 1969 and had a very strong negative influence on neural network research for over a decade. For a review of the book, see "No Harm Intended" by Jordan Pollack [66].

[29] For examples of connectionist work carried out in the 1970s and early 1980s, see Geoffrey Hinton's and James Anderson's book "Parallel Models of Associative Memory" [67].

[30] Adaptive Resonance Theory (ART) was introduced by Stephen Grossberg in 1976 and has evolved considerably since then. For a succinct summary, see the entry in *The Handbook of Brain Theory and Neural Networks* [68].

[31] Kohonen networks [69] produce topological maps in which proximate points in the input space are mapped by an unsupervised self-organizing learning process to an internal network state which preserves this topology.

Perceptron-like neural networks underwent a strong resurgence in the mid-1980s with the development of the parallel distributed processing (PDP) architecture,[32] in general, and with the introduction of the back-propagation algorithm by David Rumelhart, Geoffrey Hinton, and Ronald Williams, in particular.[33] The back-propagation learning algorithm, also known as the generalized delta rule or GDR since it is a generalization of the Widrow-Hoff delta rule for training Adaline units, overcame the limitation identified by Minsky and Papert by allowing the connection weights between the input units and the hidden units be modified, thereby enabling multi-layer perceptrons to learn solutions to problems that are not linearly separable. This was a major breakthrough in neural network and connectionist research. In cognitive science, PDP had a significant impact in promoting a move away from the sequential view of computational models of mind, towards a view of concurrently-operating networks of mutually-cooperating and competing units. PDP also played an important role in raising an awareness of the importance of the structure of the computing system on the computation, thereby challenging the functionalist doctrine of cognitivism and the principled divorce of computation from computational platform.

The standard PDP model represents a static mapping between the input vectors as a consequence of the feed-forward configuration, i.e. a configuration in which data flows in just one direction through the network, from input to output. There is an alternative, however, in which the network has connections that loop back to form circuits, *i.e.* networks in which either the output or the hidden unit activation signals are fed back to the network as inputs. These are called recurrent neural networks. The recurrent pathways in the network introduce a dynamic behaviour into the network operation.[34] Perhaps the best known type of recurrent network is the Hopfield network. These are fully recurrent networks that act as an auto-associative memory or content-addressable memory.

As a brief aside, associative memory comes in two types: hetero-associative memory and auto-associative memory. Hetero-associative memory produces an output that is different in char-

[32] David Rumelhart's and James McClelland's 1986 book *Parallel Distributed Processing: Explorations in the Microstructure of Cognition* [70] had a major influence on connectionist models of cognition.

[33] Although the back-propagation learning rule made its great impact through the work of David Rumelhart *et al.* [71, 72], it had previously been derived independently by Paul Werbos [73], among others [50].

[34] This recurrent feed-back has nothing to do with the feed-back of error signals (by, for example, back-propagation) to effect weight adjustment during learning.

acter from the input; the two are associated. Technically, the spaces to which the input and output vectors belong are different. For example, the input space might be an image of an object and the output might be a digitally-synthesized speech signal encoding a word or phrase describing the object's identity. On the other hand, auto-associative memory produces an output vector that belongs to the same space as the input vector. For example, a poorly-taken image of the object might produce — recall — a perfect image of the object taken previously.

Other recurrent networks include Elman networks (with recurrent connections from the hidden to the input units) and Jordan networks (with recurrent connections from the output to the input units). Boltzmann machines are variants of Hopfield nets that use stochastic rather than deterministic weight update procedures to avoid problems with the network becoming trapped in local minima during learning.[35]

Multi-layer perceptrons and other PDP connectionist networks typically use monotonic functions[36] such as hard-limiting threshold functions or sigmoid functions to trigger the activation of individual neurons. The use of non-monotonic activation functions, such as the Gaussian function, can offer computational advantages, *e.g.* faster and more reliable convergence on problems that are not linearly separable. Radial basis function (RBF) networks[37] use Gaussian functions but differ from multi-layer perceptrons in that the Gaussian function is used only for the hidden layer, with the input and output layers using linear activation functions.

Connectionist systems still continue to have a strong influence on cognitive science, either in a strictly PDP sense such as James McClelland's and Timothy Rogers' PDP approach to semantic cognition or in the guise of hybrid systems such as Paul Smolensky's and Geraldine Legendre's connectionist/symbolic computational architecture for cognition.[38]

With that all-too-brief overview of connectionism in mind, we can now see why connectionism, as a component of the emergent paradigm of cognitive science, is viewed as a viable and attractive alternative to cognitivism. Specifically, one of the original motivations for work on emergent systems was disaffection with

[35] For more information on Hopfield networks, Elman networks, Jordan networks, and Boltzmann machines, refer to [74], [75], [76], and [77], respectively.

[36] Monotonic functions grow in one direction only: monotonically-increasing functions only increase in value as the independent variable gets larger whereas monotonically-decreasing functions only decrease in value as the independent variable gets larger.

[37] For more details on radial basis function (RBF) networks, see for example [78].

[38] See [79] for details of James McClelland's and Timothy Rogers' PDP approach to semantic cognition and [80, 81] for details of Paul Smolensky's and Geraldine Legendre's connectionist/symbolic computational architecture for cognition.

the sequential, atemporal, and localized character of symbol-manipulation based cognitivism. Emergent systems, on the other hand, depend on parallel, real-time, and distributed architectures, just like natural biological systems do, and connectionist neural networks with their inherent capacity for learning are an obvious and appealing way to realize such systems. Of itself, however, this shift in emphasis isn't sufficient to constitute a new paradigm. While parallel distributed processing and real-time operation are certainly typical characteristics of connectionist systems, there must be more to it than this since modern cognitivist systems exhibit the very same attributes.[39] So, what are the key differentiating features? We defer answering this question until in Section 2.4, where we will compare and contrast the cognitivist and emergent paradigms on the basis of fourteen distinct characteristics. For now, we move on to consider dynamical systems approaches to emergent cognitive science.

2.2.2   Dynamical Systems

While connectionist systems focus on the pattern of activity that emerges from an adaptive network of relatively simple processing elements, dynamical systems theory models the behaviour of systems by using differential equations to capture how certain important variables that characterize the state of the system change with time. Dynamical systems theory is a very general approach and has been used to model many different types of systems in various domains such as biology, astronomy, ecology, economonics, physics, and many others.[40]

A dynamical system defines a particular pattern of behaviour. The system is characterized by a state vector and its time derivative, i.e. how it changes as time passes. This time derivative is determined by the state vector itself and also some other variables called control parameters. Usually, the dynamical equations also takes noise into account. To model a dynamical system, you need to identify the state variables and the control parameters, how noise will be modelled, and finally the exact form of the relationship which combines these and expresses them in terms of derivatives, i.e. how they change with time.

[39] Walter Freeman and Rafael Núñez have argued that recent systems — what they term neo-cognitivist systems — exploit parallel and distributed computing in the form of artificial neural networks and associative memories but, nonetheless, still adhere to many of the original cognitivist assumptions [36]. A similar point is made by Timothy van Gelder and Robert Port [82].

[40] For an intuitive introduction to dynamical systems theory, see Section 5.2 of Lawrence Shapiro's book *Embodied Cognition* [83]. For an overview of the way dynamical systems theory can be used to model cognitive behaviour, refer to Scott Kelso's book *Dynamic Patterns – The Self-Organization of Brain and Behaviour* [21].

The Nature of Dynamical Systems

In general, a dynamical system has several key attributes. First of all, it is a system. This may seem to be a bit obvious but it's important. It means that it comprises a large number of interacting components and therefore a large number of degrees of freedom.

Second, the system is dissipative, that is, it uses up or dissipates energy. This has an important consequence on the system behaviour. In particular, it means that the number of states that the system can reach reduces with time. Technically, we say that its phase space decreases in volume. The main upshot of this is that the system develops a preference for certain sets of states (again, technically, they are preferential sub-spaces in the complete space of possible states).

A dynamical system is also what is referred to as a non-equilibrium system. This just means that it never comes to rest. It doesn't mean that it can't exhibit stable behaviour — it can — but it does mean that it is unable to maintain its structure and carry out its function without external sources of energy, material, or information. In turn, this means that, at least from an energy, material, or information perspective, the system is open, i.e. stuff can enter and exit the system. In contrast, a closed system doesn't allow anything to cross the system boundary.

A dynamical system is also non-linear. This simply means that the equations that define the differential relationship between the state variables, the control parameters, and the noise components are combined together in a multiplicative manner and not simply by weighting and adding them together. Although non-linearity might appear to be a mathematical nicety (or, more likely, a mathematical complication) this non-linearity is extremely important because it provides the basis for complex behaviour — most of the world's interesting phenomena exhibit this hard-to-model characteristic of non-linearity — but, not only that, it also means that the dissipation is not uniform and that only a small number of the system's overall degrees of freedom contribute to its behaviour. In other words, when modelling the system, we need focus only on a small number of state variables instead of having to consider every single one (which would more or less make the task of modelling the system impossible). We refer to these special variables by two different, but entirely equivalent, terms: *order parameters* and *collective variables*.[41] Which term you choose is largely a matter of convention or tradition.

Each collective variable plays a key role in defining the way the system's behaviour develops over time. In essence, the collective variables are the subset of the system variables that govern the system behaviour. The main consequence of the existence of these collective variables is that the system behaviour is characterized by a succession of relatively stable states: in each state the system is doing something specific and stays doing it until something happens to cause it to jump to the next relatively stable state. For this reason, we say that the states are meta-stable (stable but subject to change) and we call the local regions in the state space around them attractors (because once a behaviour gets close to one, it is attracted to stay in the vicinity of that behaviour until something significant disturbs it).

Being able to model the behaviour of the system — a system with very many variables and therefore a very high dimensional space of *possible* states — with a very small number of relevant variables — and therefore a very low dimensional space of *relevant* states — makes the modelling exercise practical and attractive, and it is one of the main characteristics that distinguish dynamical systems from connectionist systems.

Dynamical Systems and Cognition

These are all very general characteristics of dynamical systems. So, what makes them suitable for modelling cognition? To answer this question, we need to understand the perspective that advocates of dynamical systems take on cognition. Esther Thelen and Helen Smith express it the following way:[42]

> Cognition is non-symbolic, nonrepresentational and all mental activity is emergent, situated, historical, and embodied.

To this we might add that it is socially constructed so that some aspects of cognition arise from the interaction between cognitive agents, again modelled as a dynamical process. It is clear that Thelen and Smith, along with many others who subscribe to the emergent paradigm, take issue with the symbolic nature of cognitivist models and with the representationalism it encapsulates.

[41] We already met the concept of a collective variable in Chapter 1 when we discussed Scott Kelso's ideas on the different levels of abstraction which need to be considered when modelling a system.

[42] This quotation is taken from Esther Thelen's and Helen Smith's influential book *A Dynamic Systems Approach to the Development of Cognition and Action* [84].

Here we must exercise some caution in understanding their interpretation of symbolic representationalism and their assertion that this is not a true reflection of cognition.

There are two principle ways which proponents of dynamical systems models, and emergent models in general, object to symbol manipulation and representation. One is symbol manipulation in the literal sense that a computer program manipulates symbols. In other words, the objection is to the mechanism of symbolic processing: the rule-based shuffling of symbols in search of a state that satisfies the conditions defined by the goal of the system. Instead of this, proponents of dynamical systems and connectionism contend that cognitive behaviour arises as a natural consequence of the interaction of appropriately configured network of elementary components. That is, cognition is a behaviour that is a consequence of self-organization, i.e. a global pattern of activity that arises because of, and only because of, the dynamic interaction of these components. It is emergent in the sense that this global pattern of activity cannot be predicted from the local properties of the system components.[43]

The second aspect of the objections of proponents of dynamical systems concerns the issue of representation. This is a very hotly debated issue and there is considerable disagreement over exactly what different people mean by representation.[44] As we have seen, cognitivism hinges upon the direct denotation of an object or an event in the external world by a symbolic representation that is manipulated by the cognitive system. It is this strong denotational characterization that emergent systems people object to because it entails (i.e. it necessarily involves) a correspondence between the object as it appears and is represented by the symbol and the object as it is in the world. Furthermore, by virtue of the computational functionalism of cognition, these denoted symbolic object correspondences are shared by every cognitive agent, irrespective of the manner in which the system is realized in a cognitive agent: as a computer or as a brain. This is simultaneously the power of cognitivism and a great bone of contention among those who advocate an emergent position.

So, how do emergent systems manage without representations, as the quotation above suggests they do? Here again, we

[43] Strictly speaking, the pattern of activity that arises from self-organization can *in principle* be predicted from the properties of the components so in a sense emergence is a stronger — and more obscure — process which may, or may not, exploit self-organization; see the article "Self-Organizing Systems" by Scott Camazine in *Encyclopedia of Cognitive Science* [85].

[44] We discuss the troublesome issue of representation in some depth in Chapter 8.

need to be careful in our interpretation of the term representation. It is clearly evident from what we have discussed so far that connectionist and dynamical systems exhibit different states. Could these states not be interpreted as "representing" objects and events in the world and, if so, doesn't that contradict the anti-representational position articulated above? The answer to these two questions is a conditional "yes" and a cautious "no." Such states could be construed as a representation, but not in the sense that they *denote* the object or event in the cognitivist sense. Rather, it is a question of them being correlated in some way with these objects and events but they need not mean the same thing: it's a marriage of convenience, not one of absolute commitment.

We say that such a representation *connotes* the objects or events and, in so saying, we imply nothing at all about the nature of the object or the event except that the emergent system's state is correlated in some way with its occurence.[45] If this seems to be a very fine, almost pedantic, point, it's because it is. But it is a fundamentally important point nonetheless since it goes straight to the heart of one of the core differences between the cognitivist and emergent paradigms: the relationship between the state of the agent — cognitivist or emergent — and the world it interacts with.

Cognitivism asserts that the symbolic knowledge it represents about the world is a faithful counterpart of the world itself; emergent approaches make no such claim and, on the contrary, simply allows that the internal state reflects some regularity or lawfulness in the world which it doesn't know but which it can adapt to and exploit through its dynamically-determined behaviour. This helps explain what is meant in the quotation above by cognition being *situated*, *historical*, and *embodied*. A dynamical system must be embodied in some physical way in order to interact with the world and the exact form of that embodiment makes a difference to the manner in which the agent behaves (or can behave). Being situated means that the agent's cognitive understanding of the world around it emerges in the specific context of its local surroundings, not in any virtual or abstract sense. Furthermore, the history of these context-

[45] For a deep, if also very dense, discussion of the difference between denotation and connotation in the specific context of language, refer to Alexander Kravchenko's paper "Essential properties of language, or, why language is not a code" [86].

specific interactions have an effect on the way the dynamical system develops as it continually adjusts and adapts.[46]

[46] We discuss the issue of situated embodiment in some detail in Chapter 5.

TIME

This brings us to a crucial aspect of dynamical systems which is rather obvious once we say it: it's about time.[47] To be dynamic means to change with time so it is clear that time must play a crucial part in any dynamical system. With emergent systems in general, and dynamical systems in particular, cognitive processes unfold over time. More significantly, they do so not just in an arbitrary sequence of steps, where the actual time taken to complete each step doesn't have any influence on the outcome of that step, but in real-time — in lock-step, synchronously — with events as they unfold in the world around the agent. So, time, and timing, is at the very heart of cognition and this is one of the reasons why dynamical systems theory may be an appropriate way to model it.

[47] *It's about Time: An Overview of the Dynamical Approach to Cognition* is the title of a book by Robert Port and Timothy van Gelder which is devoted to discussing the importance of time in cognition and arguing the case for a dynamical approach to modelling cognition.

The synchronicity of a dynamical cognitive agent with the events in its environment has two unexpected and, from the perspective of artificial cognitive systems, somewhat unwelcome consequences. First, it places a strong limitation on the rate at which the development of the cognitive agent can proceed. Specifically, it is constrained by the rate at which events in the world unfold and not on the speed at which internal changes can occur in the agent.[48] Biological cognitive systems have a learning cycle measured in weeks, months, and years. While it might be possible to collapse it into minutes and hours for an artificial system because of increases in the rate of internal adaptation and change, it cannot be reduced below the time-scale of the interaction. Second, taken together with the requirement for embodiment, we see that the historical and situated nature of the systems means that we cannot by-pass the developmental process: development is an integral part of cognition, at least in the emergent paradigm of cognitive science.

[48] Terry Winograd and Fernando Flores explain in their book *Computers and Cognition* [87] the impact of real-time interaction between a cognitive system and its environment on the rate at which the system can develop.

DYNAMICAL SYSTEMS AND CONNECTIONISM

We have already mentioned that there is a natural relationship between dynamical systems and connectionist systems. To a significant extent, you can consider them to be complementary ways of describing cognitive systems, with dynamical systems focussing on macroscopic behaviour and connectionist systems focussing on microscopic behaviour.[49] Connectionist systems themselves are, after all, dynamical systems with temporal properties and structures such as attractors, instabilities, and transitions. Typically, however, connectionist systems describe the dynamics in a high dimensional space of computing element activation and network connection strengths. On the other hand, dynamical systems theory describes the dynamics in a low dimensional space because a small number of state variables are capable of capturing the behaviour of the system as a whole.[50] Much of the power of dynamical perspectives comes from this higher-level abstraction of the dynamics[51] and, as we have already noted above, this is the key advantage of the dynamical systems formulation of system dynamics: it collapses a very high-dimensional system defined by the complete set of system variables onto a low-dimensional space defined by the collective variables.

The complementary nature of dynamical systems and connectionist descriptions is reflected in the approach to modelling that we met in Chapter 1 in which systems are modelled simultaneously at three distinct levels of abstraction: a boundary constraint level that determines the task or goals (initial conditions, non-specific conditions), a collective variables level which characterize coordinated states, and a component level which forms the realized system (*e.g.* nonlinearly coupled oscillators or neural networks). This complementary perspective of dynamical systems theory and connectionism enables the investigation of the emergent dynamical properties of connectionist systems in terms of attractors, meta-stability, and state transition, all of which arise from the underlying mechanistic dynamics. It also offers the possibility of implementing dynamical systems theory models with connectionist architectures.

THE STRENGTH OF THE DYNAMICAL SYSTEMS APPROACH

Those who advocate the use of dynamical systems theory to model cognition point to the fact that they provide you directly

[49] The intimate relationship between connectionism and dynamical systems is teased out in the book *Toward a New Grand Theory of Development? Connectionism and Dynamic Systems Theory Re-Considered*, edited by John Spencer, Michael Thomas, and James McClelland [88].

[50] Gregor Schöner argues that it is possible for a dynamical system model to capture the behaviour of the system using a small number of variables because the macroscopic states of high-dimensional dynamics and their long-term evolution are captured by the dynamics in that part of the space where instabilities occur: the low-dimensional Center-Manifold [89].

[51] There is a useful overview of the dynamical perspective on neural networks in a book *Mathematical perspectives on neural networks* edited by Paul Smolensky, Michael Mozer, and David Rumelhart [90]. The same book also provides useful overviews from computational and statistical viewpoints.

with many of the characteristics inherent in natural cognitive systems such as multistability, adaptability, pattern formation and recognition, intentionality, and learning. These are all achieved purely as a function of dynamical laws and the self-organization of the system that these laws enable. They require no recourse to symbolic representations, especially representations that are the result of human design: there is just an on-going process of dynamic change and formation of meta-stable patterns of system activity. They also argue that dynamical systems models allow for the development of higher order cognitive functions, such as intentionality and learning, in a relatively straight-forward manner, at least in principle. For example, intentionality — purposive or goal-directed behaviour — can be achieved by superimposing a function that encapsulates the intention onto the equations that define the dynamical system. Similarly, learning can be viewed as a modification of existing behavioural patterns by introducing changes that allow new meta-stable behaviours to emerge, i.e. by developing new attractors in the state space. These changes don't just add the extra meta-stable patterns but in doing so they also may have an effect on existing attractors and existing behaviour. Thus, learning changes the whole system as a matter of course.

While dynamical models can account for several non-trivial behaviours that require sensorimotor learning and the integration of visual sensing and motoric control (e.g. the perception of affordances,[52] perception of time to contact,[53] and figure-ground bi-stability[54] the feasibility of realizing higher-order cognitive faculties has not yet been demonstrated. It appears that dynamical systems theory (and connectionism) needs to be embedded in a larger emergent context. This makes sense when you consider dynamical systems theory and connectionism from the ultimate-proximate perspective we discussed in Chapter 1: both focus more on the proximate aspects of modelling methodology and mechanisms of cognition than on the ultimate concern of what cognition is for, an issue to which we now turn.

[52] The concept of *affordance* is due to the influential psychologist J. J. Gibson [91]. It refers to the potential use to which an object can be put, *as perceived by an observer* of the object. This perceived potential depends on the skills the observer possesses. Thus, the affordance is dependent both on the object itself and the perception and action capabilities of the observing agent.

[53] The *time-to-contact* refers to the time remaining before an agent or a part of the agent's body will make contact with something in the agent's environment. It is often inferred from optical flow (a measure of how each point in the visual field of an agent is moving) and is essential to many behaviours; e.g., as Scott Kelso illustrates in his book *Dynamic Patterns*, gannets use optical flow to determine when to fold their wings before entering the water as they dive for fish [21].

[54] *Figure-ground bi-stability* refers to the way we alternately see one shape (the figure) or another (the background, formed by the complement of the figure), but never both at the same time; see Wolfgang Köhler's *Dynamics in Psychology* [92] for more details.

2.2.3 *Enactive Systems*

We now focus on an increasingly-important approach in cognitive science: enaction.[55] The principal idea of enaction is that a cognitive system develops it own understanding of the world around it through its interactions with the environment. Thus, enactive cognitive system operate autonomously and generate their own models of how the world works.

THE FIVE ASPECTS OF ENACTION

When dealing with enactive systems, there are five key elements to consider. These are:

1. Autonomy

2. Embodiment

3. Emergence

4. Experience

5. Sense-making

We have already encountered the first four of these elements.

The issue of autonomy was introduced in Chapter 1, Section 1.3, where we noted the link between cognition and autonomy, particularly from the perspective of biological systems. We take up this important issue again later in the book and devote all of Chapter 4 to unwrapping the somewhat complex relationship between the two topics.

Similarly, in this chapter we have already met the concept of embodiment and the related concept of embodied cognition. Again, a full chapter is dedicated to embodiment later in the book (Chapter 5) reflecting is importance in contemporary cognitive science.

Emergence is, of course, the topic of the current section and we have already discussed the relationship between emergence and self-organization (see Section 2.2.2). Emergence refers to the phenomenon whereby the behaviour we call cognition arises from the dynamic interplay between the components of the system and between the components and the system as a whole. We return to this issue in Chapter 4, Section 4.3.5.

[55] Section 2.2.3 is based directly on a study by the author of enaction as a framework for development in cognitive robotics [93]. The paper contains additional technical details relating to enactive systems which are not strictly required here. Readers who are interested in delving more deeply into enaction are encouraged to refer to this paper as well as to the original literature [14, 32, 48, 87, 94, 95, 96, 97, 98]. The book *The Embodied Mind* by Francisco Varela, Evan Thompson, and Eleanor Rosch [98] would make a good starting point, followed perhaps by the book *Enaction: Toward a New Paradigm for Cognitive Science* by John Stewart, Olivier Gapenne, and Ezequiel Di Paolo [48] for a contemporary perspective on Enaction.

Experience is the fourth element of enaction and, as we noted in the introduction to this section, it is simply the cognitive system's history of interaction with the world around it: the actions it takes in the environment in which it is embedded and the actions arising in the environment which impinge on the cognitive system. In enactive systems, these interactions don't control the system, otherwise it wouldn't be autonomous and, notwithstanding what we said in Chapter 1 about having to be cautious in approaching the relationship between cognition and autonomy, enactive systems are by definition autonomous. Even so, these interactions can and do trigger changes in the state of the system. The changes that can be triggered are *structurally determined*: they depend on the system structure, *i.e.* the embodiment of the self-organizational principles that make the system autonomous.[56] This structure is also referred to as the system's *phylogeny*: the innate capabilities of an autonomous system with which it is equipped at the outset (when it is born, in the case of a biological system) and which form the basis for its continued existence. The experience of the system — its history of interactions — involving *structural coupling*[57] between the system and its environment in an ongoing process of mutual perturbation is referred to as its *ontogeny*.

Finally, we come to the fifth and, arguably, the most important element of enaction: sense-making. This term refers to the relationship between the knowledge encapsulated by a cognitive system and the interactions which gave rise to it. In particular, it refers to the idea that this emergent knowledge is generated by the system itself and that it captures some regularity or lawfulness in the interactions of the system, *i.e.* its experience. However, the sense it makes is dependent on the way in which it can interact: its own actions and its perceptions of the environment's action on it. Since these perceptions and actions are the result of an emergent dynamic process that is first and foremost concerned with maintaining the autonomy and operational identity of the system, these perceptions and actions are unique to the system itself and the resultant knowledge makes sense only insofar as it contributes to the maintenance of the system's autonomy. This ties in neatly with the view of cognition as a pro-

[56] The founders of the enactive approach use the term *structural determination* to denote the dependence of a system's space of viable environmentally-triggered changes on the structure and its internal dynamics [14, 98]. The interactions of this structurally-determined system with the environment in which it is embedded are referred to as *structural coupling*: a process of mutual perturbations of the system and environment that facilitate the on-going operational identity of the system and its autonomous self-maintenance. Furthermore, the process of structural coupling produces a congruence between the system and its environment. For this reason, we say that the system and the environment are *co-determined*. The concepts of structural determination and structural coupling of autopoietic systems [14] are similar to Scott Kelso's circular causality of action and perception [21] and the organizational principles inherent in Mark Bickhard's self-maintenant systems [13]. The concept of enactive development has its roots in the structural coupling of organizationally-closed systems which have a central nervous system and is mirrored in Bickhard's concept of recursive self-maintenance [13].
[57] *Structural coupling*: see Sidenote 56 above.

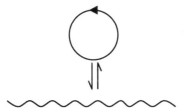

Figure 2.3: Maturana and Varela's ideogram to denote a structurally-determined organizationally-closed system. The arrow circle denotes the autonomy and self-organization of the system, the rippled line the environment, and the bi-directional half-lines the mutual perturbation — structural coupling — between the two.

cess that anticipates events and increases the space of actions in which a system can engage.

By making sense of its experience, the cognitive system is constructing a model that has some predictive value, exactly because it captures some regularity or lawfulness in its interactions. This self-generated model of the system's experience lends the system greater flexibility in how it interacts in the future. In other words, it endows the system with a larger repertoire of possible actions that allow richer interactions, increased perceptual capacity, and the possibility of constructing even better models that encapsulate knowledge with even greater predictive power. And so it goes, in a virtuous circle. Note that this sense-making and the resultant knowledge says nothing at all about what is really out there in the environment. It doesn't have to: all it has to do is make sense for the continued existence and autonomy of the cognitive system.

Sense-making is actually the source of the term enaction. In making sense of its experience, the cognitive system is somehow bringing out through its actions — enacting — what is important for the continued existence of the system. This enaction is effected by the system as it is embedded in its environment, but as an autonomous entity distinct from the environment, through an emergent process of making sense of its experience. To a large extent, this process of sense-making is exactly what we mean by cognition (in the emergent paradigm, at least).

Enaction and Development

The founders of the enactive approach, Humberto Maturana and Francisco Varela, introduced a diagrammatic way of conveying the self-organizing and self-maintaining autonomous nature of an enactive system, perturbing and being perturbed by its environment: see Figure 2.3.[58] The arrowed circle denotes the autonomy and self-organization of the system, the rippled line the environment, and the bi-directional half-arrows the mutual perturbation.

We remarked above that the process of sense-making forms a virtuous circle in that the self-generated model of the system's experience provides a larger repertoire of possible actions, richer interactions, increased perceptual capacity, and potentially better self-generated models, and so on. Recall also our earlier remarks that the cognitive system's knowledge is represented by the state of the system. When this state is embodied in the system's central nervous system, the system has much greater plasticity in two senses: (a) the nervous system can accommodate a much larger space of possible associations between system-environment interactions, and (b) it can accommodate a much larger space of potential actions. Consequently, the process of cognition involves the system modifying its own state, specifically its central nervous system, as it enhances its predictive capacity and its action capabilities. This is exactly what we mean by development. This generative (*i.e.* self-constructed) autonomous learning and development is one of the hallmarks of the enactive approach.

Development is the cognitive process of establishing and enlarging the possible space of mutually-consistent couplings in which a system can engage (or, perhaps more appropriately, which it can withstand without compromising its autonomy). The space of perceptual possibilities is founded not on an absolute objective environment, but on the space of possible actions that the system can engage in while still maintaining the consistency of the coupling with the environment. These environmental perturbations don't control the system since they are not components of the system (and, by definition, don't play a part in the self-organization) but they do play a part in the ontogenetic development of the system. Through this ontogenetic development, the cognitive system develops its own epistemology, *i.e.* its own system-specific history- and context-dependent knowledge of its

[58] The Maturana and Varela ideograms depicting self-organizing, self-maintaining, developmental systems appear in their book *The Tree of Knowledge — The Biological Roots of Human Understanding* [14].

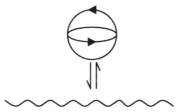

Figure 2.4: Maturana and Varela's ideogram to denote a structurally-determined organizationally-closed autonomous system *with a central nervous system*. This system is capable of development by means of self-modification of its nervous system, so that it can accommodate a much larger space of effective system action.

world, knowledge that has meaning exactly because it captures the consistency and invariance that emerges from the dynamic self-organization in the face of environmental coupling. Again, it comes down to the preservation of autonomy, but this time doing so in an ever-increasing space of autonomy-preserving couplings.

This process of development is achieved through self-modification by virtue of the presence of a central nervous system: not only does environment perturb the system (and *vice versa*) but the system also perturbs itself and the central nervous system adapts as a result. Consequently, the system can develop to accommodate a much larger space of effective system action. This is captured in a second ideogram of Maturana and Varela (see Figure 2.4) which adds a second arrow circle to the ideogram to depict the process of development through self-perturbation and self-modification. In essence, development *is* autonomous self-modification and requires the existence of a viable phylogeny, including a nervous system, and a suitable ontogeny.

Knowledge and Interaction

Let us now move on to discuss in a little more detail the nature of the knowledge that an enactive cognitive system constructs. This knowledge is built on sensorimotor associations, achieved initially by exploration of what the world offers. However, this is only the beginning. The enactive system uses the knowledge gained to form new knowledge which is then subjected to empirical validation to see whether or not it is warranted. After all, we, as enactive beings, imagine many things but not everything we imagine is valid in the sense that it is plausible or

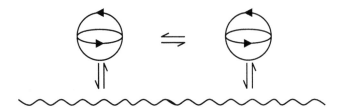

Figure 2.5: Maturana and Varela's ideogram to denote the development engendered by interaction between cognitive systems

corresponds well with reality. This brings us to one of the key issues in cognition: internal simulation, i.e. the ability to rehearse a train of imagined perceptions and actions, and assess the likely outcome in memory. This internal simulation is used to build on sensorimotor knowledge and accelerate development. Internal simulation thus provides the key characteristic of cognition: the ability to predict future events, to reconstruct or explain observed events (constructing a causal chain leading to that event), or to imagine new events.[59] Naturally, there is a need to focus on (re-)grounding predicted, explained, or imagined events in experience so that the system can *do* something new and interact with the environment in a new way. If the cognitive system wishes or needs to share this knowledge with other cognitive systems or communicate with other cognitive systems, it will only be possible if they have shared a common history of experiences and if they have a similar phylogeny and a compatible ontogeny. In essence, *the meaning of the knowledge that is shared is negotiated and agreed by consensus through interaction.*

When there are two or more cognitive agents involved, interaction is a shared activity in which the actions of each agent influence the actions of the others, resulting in a mutually constructed pattern of shared behaviour. Again, Humberto Maturana and Francisco Varela introduce a succinct diagrammatic way of of conveying this coupling between cognitive agent and the development it engenders: see Figure 2.5.[60] Thus, explicit meaning is not necessary for something to be communicated in an interaction, it is simply necessary that the agents are engaged in a mutual sequence of actions. Meaning emerges through shared consensual experience mediated by interaction.

[59] For more details on the nature of internal simulation, see [99, 100, 101]. We return to this topic in Sections 5.8 and 7.5.

[60] Such mutually-constructed patterns of complementary behaviour is also emphasized in Andy Clark's notion of joint action [102].

SUMMARY

To recap: enaction involves two complementary processes: (a) phylogenetically-dependent structural determination, *i.e.* the preservation of autonomy by a process of self-organization which determines the relevance and meaning of the system's interactions, and (b) ontogenesis, *i.e.* the increase in the system's predictive capacity and the enlargement of its action repertoire through a process of model construction by which the system develops its understanding of the world in which is it embedded. Ontogenesis results in development: the generation of new couplings effected by the self-modification of the system's own state, specifically its central nervous system. This complementarity of structural determination — phylogeny — and development — ontogeny — is crucial.

Cognition is the result of a developmental process through which the system becomes progressively more skilled and acquires the ability to understand events, contexts, and actions, initially dealing with immediate situations and increasingly acquiring a predictive or prospective capability. Prediction, or anticipation, is one of the two hallmarks of cognition, the second being the ability to learn new knowledge by making sense of its interactions with the world around it and, in the process, enlarging its repertoire of effective actions. Both anticipation and sense-making are the direct result of the developmental process. This dependency on exploration and development is one of the reasons why an artificial cognitive system requires a rich sensory-motor interface with its environment and why embodiment plays such a pivotal role.

2.3   Hybrid Systems

Cognitivist and emergent paradigms of cognitive science clearly have very different outlooks on cognition and they each have their own particular strengths and weaknesses. Thus, it would seem to be a good idea to combine them in a hybrid system that tries to get the benefits of both without the disadvantages of either. This is what many people try to do. Typically, hybrid systems exploit symbolic knowledge to represent the agent's world and logical rule-based systems to reason with this knowledge to pursue tasks and achieve goals. At the same time, they typically use emergent models of perception and action to explore the world and construct this knowledge. While hybrid systems still use symbolic representations, the key idea is that they are constructed by the system itself as it interacts with and explores the world. So, instead of a designer programming in all the necessary knowledge, objects and events in the world can be represented by observed correspondences between sensed perceptions, agent actions, and sensed outcomes. Thus, just like an emergent system, a hybrid system's ability to understand the external world is dependent on its ability to flexibly interact with it. Interaction becomes an organizing mechanism that establishes a learned association between perception and action.

2.4   A Comparison of Cognitivist and Emergent Approaches

Although cognitivist and emergent approaches are often contrasted purely on the basis of their use of symbolic representation — or not, as the case may be — it would be a mistake to think that this is the only issue on which they differ and, equally, it would be wrong to assume that the distinction is as black-and-white as it is sometimes presented. As we have seen, symbols have a place in both paradigms; the real issue is whether these symbols denote things in the real world or simply connote them from the agent's perspective. In fact, we can contrast the cognitivist and emergent paradigms in many different ways. The following are fourteen characteristics that have proven to be useful in drawing out the finer distinctions between the two paradigms.[61]

1. Computational operation
2. Representational framework
3. Semantic grounding
4. Temporal constraints
5. Inter-agent epistemology
6. Embodiment
7. Perception

[61] These fourteen characteristics are based on the twelve proposed by the author, Giorgio Metta, and Giulio Sandini in a paper entitled "A Survey of Artificial Cognitive Systems: Implications for the Autonomous Development of Mental Capabilities in Computational Agents" [103]. They have been augmented here by adding two more: the role of cognition and the underlying philosophy. The subsequent discussion is also an extended version of the commentary in [103].

8. Action

9. Anticipation

10. Adaptation

11. Motivation

12. Autonomy

13. The role of cognition

14. Philosophical foundations

Let us look at each of these in turn. In doing so, we sometimes have to refer to concepts that are covered later in the book. The relevant chapters or sections or end-notes are indicated in the end-notes.

Computational operation: cognitivist systems use rule-based manipulation of symbol tokens, typically but not necessarily in a sequential manner. On the other hand, emergent systems exploit processes of self-organization, self-production, self-maintenance, and development, through the concurrent interaction of a network of distributed interacting components.

Representational framework: cognitivist systems use patterns of symbol tokens that denote events in the external world. These often describe how the designer sees the relationship between the representation and the real-world, the assumption being that all agents see the world in the same way. The representations of emergent systems are global system states encoded in the dynamic organization of the system's distributed network of components.

Semantic grounding: semantic representations reflect the way that a particular cognitive agent sees the world. Cognitivist systems ground symbolic representations by identifying percepts with symbols, either by design or by learned association. These representations are accessible to direct human interpretation. In contrast, emergent systems ground representations by autonomy-preserving anticipatory and adaptive skill construction. These representations only have meaning insofar as they contribute to the continued viability of the system and are inaccessible to direct human interpretation.

Temporal constraints: cognitivist systems operate atemporally in the sense that time is not an essential element of the computation. It is just a measure of how long it takes to get the result and these results won't change with the amount of time taken. However, emergent systems are entrained by external events and timing is an intrinsic aspect of how they operate. The timing of the system's behaviour relative to the world's behaviour is crucial. This also limits the speed with which they can learn and develop.

Inter-agent epistemology: for cognitivist systems, an absolute shared epistemology, i.e. framework of knowledge, between agents is guaranteed by virtue of their positivist stance on reality; that is, each agent is embedded in an environment, the structure and semantics of which are independent of the system's cognition. This contrasts strongly with emergent systems for which epistemology is the subjective agent-specific outcome of a history of shared consensual experiences among phylogentically-compatible agents. This position reflects the phenomenological stance on reality taken by emergent systems, in general, and enaction, in particular.

Embodiment: cognitivist systems do not need to be embodied, in principle, by virtue of their roots in computational functionalism which holds that cognition is independent of the physical platform in which it is implemented. Again, in contrast, emergent systems are necessarily embodied and the physical realization of the cognitive system plays a direct constitutive role in the cognitive process.

Perception: in cognitivist systems, perception provides an interface between the absolute external world and the symbolic representation of that world. The role of perception is to abstract faithful spatio-temporal representations of the external world from sensory data. In emergent systems, perception is an agent-specific interpretation of the way the environment perturbs the agent and is, at least to some extent, dependent on the embodiment of the system.

Action: in cognitivist systems, actions are causal consequences of symbolic processing of internal representations, usually carried out when pursuing some task. In emergent systems, actions are the way the agent perturbs the environment, typically to maintain the viability of the system. In both cases, actions are directed by the goals these actions are intended to fulfil.

Anticipation: in cognitivist systems, anticipation typically takes the form of planning using some form of procedural or probabilistic reasoning with some prior model. Anticipation in the emergent paradigm takes the form of the cognitive system visiting some subset of the states in its self-constructed perception-action state space but without commiting to the associated actions.

Adaptation: for cognitivism, adaptation ususally implies the acquisition of new knowledge. In emergent systems, adaptation entails a structural alteration or re-organization to effect a new set of dynamics. Adaptation can take the form of either learning or development; Chapter 6 explains the difference.

Motivation: in cognitivist systems, motives provide the criteria which are used to select a goal and the associated actions. In emergent systems, motives encapsulate the implicit value system that modulate the system dynamics of self-maintenance and self-development, impinging on perception (through attention), action (through action selection), and adaptation (through the mechanisms that govern change), such as enlarging the space of viable interaction.

Autonomy: the cognitivist paradigm does not require the cognitive agent to be autonomous but the emergent paradigm does. This is because in the emergent paradigm cognition is the process whereby an autonomous system becomes viable and effective through a spectrum of homeostatic processes of self-regulation. Chapter 4 explains the concept of homeostasis and exands further on the different nuances of autonomy.

Role of cognition: in the cognitivist paradigm, cognition is the rational process by which goals are achieved by reasoning with symbolic knowledge representations of the world in which the agent operates. This contrasts again with the emergent paradigm, in which cognition is the dynamic process by which the system acts to maintain its identity and organizational coherence in the face of environmental perturbation. Cognition entails system development to improve its anticipatory capabilities and

The Cognitivist Paradigm vs. the Emergent Paradigm		
Characteristic	Cognitivist	Emergent
Computational Operation	Syntactic manipulation of symbols	Concurrent self-organization of a network
Representational Framework	Patterns of symbol tokens	Global system states
Semantic Grounding	Percept-symbol association	Skill construction
Temporal Constraints	Atemporal	Synchronous real-time entrainment
Inter-agent epistemology	Agent-independent	Agent-dependent
Embodiment	No role implied: functionalist	Direct constitutive role: non-functionalist
Perception	Abstract symbolic representations	Perturbation by the environment
Action	Causal result of symbol manipulation	Perturbation by the system
Anticipation	Procedural or probabilistic reasoning	Traverse of perception-action state space
Adaptation	Learn new knowledge	Develop new dynamics
Motivation	Criteria for goal selection	Increase space of interaction
Autonomy	Not entailed	Cognition entails autonomy
Role of Cognition	Rational goal-achievement	Self-maintenance and self-development
Philosophical Foundation	Positivism	Phenomenology

Table 2.1: A comparison of cognitivist and emergent paradigms of cognition; refer to the text for a full explanation (adapted from [103] and extended).

extend its space of autonomy-preserving actions.

Philosophical foundations: the cognitivist paradigm is grounded in positivism, whereas the emergent paradigm is grounded in phenomenology.[62]

Table 2.1 presents a synopsis of these key issues.

2.5   *Which Paradigm Should We Choose?*

The cognitivist, emergent, and hybrid paradigms each have their proponents and their critics, their attractions and their challenges, their strong points and their weak points. However, it is crucial to appreciate that each paradigm is not equally well developed as a science and so it isn't possible to make any definitive judgement on their long-term prospects. At the same time, it is important to recognize that while the arguments in favour of emergent systems are very compelling, the current capabilities of cognitivist systems are more advanced. At present, you can do far more with a cognitivist system than an emergent one

[62] For a discussion of the positivist roots of cognitivism, see "Restoring to Cognition the Forgotten Primacy of Action, Intention and Emotion" by Walter Freeman and Rafael Núñez [36]. The paper "Enactive Artificial Intelligence: Investigating the systemic organization of life and mind" by Tom Froese and Tom Ziemke [104] discusses the phenomenological leanings of enaction. A paper by the author and Dermot Furlong [105], "Philosophical Foundations of Enactive AI," provides an overview of the philosophical traditions of AI and cognitive science.

(from the perspective of artificial cognitive systems, at any rate). With that in mind, we wrap up this chapter by looking briefly at some of their respective strengths and weaknesses, and how they might be resolved.

According to some, cognitivist systems suffer from three problems:[63] the symbol grounding problem (the need to give symbolic representations some real-world meaning; see Chapter 8, Section 8.4), the frame problem (the problem of knowing what does and does not change as a result of actions in the the world),[64] and the combinatorial explosion problem (the problem of handling the large and possibly intractable number of new relations between elements of a representation when something changes in that representation as a consequence of some action; see Sidenote 10 in this chapter). These problems are put forward as reasons why cognitivist models have difficulties in creating systems that exhibit robust sensori-motor interactions in complex, noisy, dynamic environments, and why they also have difficulties modelling the higher-order cognitive abilities such as generalization, creativity, and learning. A common criticism of cognitivist systems is that they are are poor at functioning effectively outside narrow, well-defined problem domains, typically because they depend so much on knowledge that is provided by others and that depends very often on implicit assumptions about the way things are in the world in which they are operating. However, setting aside one's scientific and philosophical convictions, this criticism of cognitivism is unduly harsh because the alternative emergent systems don't perform particularly well at present (except, perhaps, in principle).

Emergent systems should in theory be much less brittle because they emerge — and develop — through mutual specification and co-determination with the environment. However, our ability to build artificial cognitive systems based on these principles is very limited at present. To date, dynamical systems theory has provided more of a general modelling framework rather than a model of cognition and has so far been employed more as an analysis tool than as a tool for the design and synthesis of cognitive systems. The extent to which this will change, and the speed with which it will do so, is uncertain.

[63] For more details on the problems associated with cognitivism, see Wayne Christensen's and Cliff Hooker's paper "Representation and the Meaning of Life" [106].

[64] In the cognitivist paradigm, the frame problem has been expressed in slightly different but essentially equivalent terms: how can one build a program capable of inferring the effects of an action without reasoning explicitly about all its perhaps very many non-effects? [107].

Hybrid approaches appear to offer the best of both worlds: the adaptability of emergent systems (because they populate their representational frameworks through learning and experience) and also the advanced starting point of cognitivist systems (because the representational invariances and representational frameworks don't have to be learned but are designed in). However, it is unclear how well one can combine what are ultimately highly antagonistic underlying philosophies. Opinion is divided, with arguments both for and against.[65] One possible way forward is the development of a form of *dynamic computationalism* in which dynamical elements form part of an information-processing system.[66]

Clearly, there are some fundamental differences these two general paradigms — for example, the principled body-independent nature of cognitivist systems vs. the body-dependence of emergent developmental systems, and the manner in which cognitivist systems often preempt development by embedding externally-derived domain knowledge and processing structures — but the gap between the two shows some signs of narrowing. This is mainly due to (i) a fairly recent recognition on the part of proponents of the cognitivist paradigm of the important role played by action and perception in the realization of a cognitive system; (ii) a move away from the view that internal symbolic representations are the only valid form of representation; and (iii) a weakening of the dependence on embedded pre-programmed knowledge and the attendant increased use of machine learning and statistical frameworks both for tuning system parameters and the acquisition of new knowledge.

Cognitivist systems still have some way to go to address the issue of true ontogenetic development with all that it entails for autonomy, embodiment, architecture plasticity, and agent-centred construction of knowledge, mediated by exploratory and social motivations and innate value systems. Nevertheless, to some extent they are moving closer together in the ultimate dimension of the ultimate-proximate space, if not in the proximate dimension. This shift is the source of the inter-paradigm resonances we mentioned in this chapter and the previous one. However, since fundamental differences remain it is highly un-

[65] For the case in favour of hybrid systems, see e.g. [33, 49, 108]; for the case against, see e.g. [106]

[66] Apart from offering a way out of the cognitivist-emergent stand-off through *dynamic computationalism*, Andy Clark's book *Mindware – An Introduction to the Philosophy of Cognitive Science* [33] provides a good introduction to the foundational assumptions upon which both paradigms are based. Regarding dynamic computationalism, James Crutchfield, whilst agreeing that dynamics are certainly involved in cognition, argues that dynamics *per se* are "not a substitute for information processing and computation in cognitive processes" [49]. He puts forward the idea that a synthesis of the two can be developed to provide an approach that does allow dynamical state space structures to support computation and he proposes *computational mechanics* as the way to tackle this synthesis of dynamics and computation.

likely they will ever fully coalesce. This puts hybrid systems in a difficult position. For them to be a real solution to the cognitivist/emergent dilemma, they need to overcome the deep-seated differences discussed in the previous section.

Let us close this chapter with a reminder that, to date, no one has designed and implemented a complete cognitive system. So, on balance, the jury is still out on which paradigm to choose as the best model of an artificial cognitive system, especially given that both fields continue to evolve. Nonetheless, we need to move forward and make some choices if we are to realize an artifical cognitive system. For cognitive science, this process of realization begins with the specification of what is known as the cognitive architecture, the subject of the next chapter.

3
Cognitive Architectures

3.1 What Is a Cognitive Architecture?

When we think of architecture, typically what comes to mind is the design of buildings that satisfy some functional need but do so in a way that appeals to the people that use them. Often, architecture inspires some sense of the extraordinary and gives an impression of cohesion that makes the building whole and self-contained. Since the architectural process involves not just imagining bold new concepts but also the creation of detailed designs and technical specifications, architecture has been borrowed by many other desciplines to serve as a catch-all term for the technical specification and design of any complex artifact. Just as with architecture in the built environment, system architecture addresses both the conceptual form and the utilitarian functional aspects of the system, focussing on inner cohesion and self-contained completeness.

We use the term cognitive architecture in exactly this way to reflect the specification of a cognitive system, its components, and the way these components are dynamically related as a whole.

One of the most famous maxims in architecture, and in design generally, is that "form follows function,"[1] the principle that the shape of a building, or any object, should be mainly based on its intended purpose or function. However, in contemporary architecture this is interpreted very broadly to include both utility and aesthetic value: the degree to which it engenders a posi-

[1] The idea that form follows function is due to the nineteenth century architect Louis Sullivan.

tive interaction between people and the building and the degree to which the building is perceived as a complete entity. As we noted in the previous chapter, interaction plays a key role in cognition (and *vice versa*) so this broad interpretation of architecture is very apt when it comes to cognitive architecture: the system the architecture describes must work both at a global system level, enabling the effective interaction of a cognitive agent with the world around it, and at a component level, showing how all the parts should fit together to create the global system: the cohesive whole.

Just as there are different styles and traditions in traditional architecture, each emphasizing different facets of form and function, so too there are many different styles of cognitive architecture, each derived, more or less directly, from the three paradigms of cognitive science we discussed in the previous chapter: the cognitivist, the emergent, and the hybrid. However, the term cognitive architecture can in fact be traced to pioneering work in cognitivist cognitive science.[2] Consequently, it means something very specific in cognitivism. In particular, a cognitive architecture represents any attempt to create what is referred to as a unified theory of cognition.[3] This is a theory that covers a broad range of cognitive issues, such as attention, memory, problem solving, decision making, and learning. Furthermore, a unified theory of cognition should cover these issues from several aspects including psychology, neuroscience, and computer science. Allen Newell's and John Laird's Soar[4] architecture, John Anderson's ACT-R[5] architecture, and Ron Sun's CLARION architecture are typical candidate unified theories of cognition.[6]

Since unified theories of cognition are concerned with the computational understanding of *human* cognition, cognitivist cognitive architectures are concerned with human cognitive science as well as artificial cognitive systems. There is an argument that the term cognitive architecture should be reserved for systems that model human cognition and that the term "agent architecture" would be a better term to refer to general intelligent behaviour, including both human and computer-based artificial cognition. However, it has become common to use the term cognitive architecture in this more general sense so we will

[2] The term cognitive architecture is due to Allen Newell and his colleagues in their work on unified theories of cognition [41, 43].

[3] Unified theories of cognition are discussed in depth in Allen Newell's book of the same name [43] and John Anderson's paper "An integrated theory of the mind" [109].

[4] For more details on the Soar cognitive architecture, please refer to the papers by Allen Newell, John Laird, and colleagues [42, 110, 111, 112, 113], John Laird's book [114], and read Section 3.4.1 in this chapter.

[5] For more details on the ACT-R cognitive architecture, please refer to [109, 115].

[6] The CLARION cognitive architecture is described in depth in, e.g., [116, 117].

use it throughout the book to refer to both human and artificial cognitive systems.

Although the term cognitive architecture originated in cognitivist cognitive science, it has also been adopted in the emergent paradigm where it has a sightly different meaning. Consequently, we will begin by considering exactly what a cognitive architecture does involve in the two different approaches: cognitivist and emergent. Following that, we will discuss the features of a cognitive architecture that are considered to be necessary and desirable. Finally, we will look at three specific cognitive architectures — one from the cognitivist paradigm of cognitive science, one from the emergent, and one from the hybrid paradigm — in different levels of detail to get some understanding of what they involve and the role they play in the design of a working cognitive system.

3.1.1 *The Cognitivist Perspective*

In the cognitivist paradigm, the focus in a cognitive architecture is on the aspects of cognition that are *constant over time* and that are *independent of the task*.[7] In the words of Ron Sun, a leading exponent of cognitive architectures, [17]:

> "a cognitive architecture is a broadly-scoped domain-generic computational cognitive model, capturing the essential structure and process of the mind, to be used for broad, multiple-level, multiple-domain analysis of behaviour."

Since a cognitive architecture represents the fixed part of cognition, it cannot accomplish anything in its own right. A cognitivist cognitive architecture is a generic computational model that is neither domain-specific nor task-specific. To do something, i.e. to perform a given task, it needs to be provided with the knowledge to perform any given task. It is the knowledge which populates the cognitive architecture that provides the means to perform a task or to behave in some particular way. This combination of a given cognitive architecture and a particular knowledge set is generally referred to as a *cognitive model*.

[7] The idea that a cognitive architecture focusses on those aspects of cognition that are constant over time and independent of the task, i.e. unchanging from situation to situation, is widely supported in the literature; for example, see [118, 119, 120, 121].

So, where does this knowledge come from? In most cognitivist systems the knowledge incorporated into the model is normally determined by the person who designed the architecture, and often this knowledge is highly crafted, possibly drawing on years of experience working in the problem domain. Machine learning is increasingly used to augment and adapt this knowledge but typically you need to provide a critical minimum amount of knowledge in order to get the learning started.

The cognitive architecture itself determines the overall structure and organization of a cognitive system, including the component parts or modules, the relations between these modules, and the essential algorithmic and representational details within them. The architecture specifies the formalisms for knowledge representations and the types of memories used to store them, the processes that act upon that knowledge, and the learning mechanisms that acquire it. Usually, it also provides a way of programming the system so that a cognitive system can be customized for some application domain.

A cognitive architecture plays an important role in computational modelling of cognition in that it makes explicit the set of assumptions upon which that cognitive model is founded. These assumptions are typically derived from several sources: biological or psychological data, philosophical arguments, or *ad hoc* working hypotheses inspired by work in different disciplines such as neurophysiology, psychology, or artificial intelligence. Once it has been created, a cognitive architecture also provides framework for developing the ideas and assumptions encapsulated in the architecture.

3.1.2 *The Emergent Perspective*

Emergent approaches to cognition focus on the development of the agent from a primitive state to a fully cognitive state over its life-time. Although the concept of a cognitive architecture has its origins in cognitivism as the timeless fixed part of a cognitive system that provides the framework for processing knowledge, the term cognitive architecture is also used with emergent approaches. In this case, it isn't so much the framework that complements the knowledge as it is the framework that facilitates development. In this sense, an emergent cognitive architecture is essentially equivalent to the phylogenetic configuration of a new-born cognitive agent: the initial state from which it subsequently develops. In other words, an emergent cognitive architecture is everything a cognitive system needs to get started. This doesn't guarantee successful development, though, because development also requires exposure to an environment that is conducive to development, one in which there is sufficient regularity to allow the system to build a sense of understanding of the world around it, but not excessive variety that would overwhelm an agent which has inherent limitations on the speed with which it can develop. Thus, in a way that parallels the two-sided coin of cognitivist cognition — architecture and knowledge — emergent cognition also has two sides: architecture and gradually-acquired experience. These two sides of the emergent coin are referred to as phyogeny and ontogeny (or ontogenesis), the latter being the interactions and experiences that a developing cognitive system is exposed to as it acquires an increasing degree of cognitive capability.

With emergent approaches, the cognitive architecture provides a way of dealing with the intrinsic complexity of a cognitive system, by providing some form of structure within which to embed the mechanisms for perception, action, adaptation, anticipation, and motivation that enable the ontogenetic development over the system's life-time. It is this complexity that distinguishes an emergent developmental cognitive system from, for example, a connectionist system such as an artificial neural network that performs just one or two functions such as recognition or control. Of course, an emergent cognitive architecture might comprise many individual neural networks and, as we will see later, some do.

So, the cognitive architecture of an emergent system thus provides the basis for its subsequent development. It's worth remarking that, as a consequence of this development, the architecture itself might change. Thus, an emergent cognitive architecture isn't necessarily fixed and timeless: it is a point of departure.

The presence of innate capabilities in an emergent system does not imply that the architecture is necessarily functionally modular, *i.e.* that the cognitive system is comprised of distinct modules each one carrying out a specialized cognitive task.[8] If modularity is present, it may be because it develops this modularity through experience as part of its ontogenesis rather than being prefigured by the phylogeny of the system. The cognitivist and emergent perspectives differ somewhat on the issue of innate structure. While in an emergent system the cognitive architecture *is* the innate structure, this is not necessarily so with a cognitivist system.[9]

Sometimes, especially in developmental robotics, the term epigenesis is used instead of ontogensis, and developmental robotics is sometimes referred to as epigenetic robotics. Epigenesis has its roots in biology where it refers to the way an organism develops through cell-division into a a viable complex entity. This happens through gene expression so that the epigenesis refers to the changes that result from factors other than those determined by the organism's DNA. Ontogenesis also refers to early cellular development but more generally it refers to the development of the organism *over its full lifetime*. Thus, it includes the development of the agent after birth, including its cognitive development, and so embraces, for example, developmental psychology. Since the epigenetic process focusses exclusively on the very early growth of the agent and the way its final structure is determined, in artificial cognitive systems, epigenesis would probably be better reserved to reflect the autonomous formation and construction of cognitive architecture prior to development as a consequence of experience. To avoid confusion, we will avoid using the term epigenesis and epigenetic robotics, and refer to ontogeneis and developmental robotics on the understanding that we are discussing the development of an entity after it has been born (in the case of natural cognitive systems) or realized as a physical system (in the case of artifical cognitive systems). For the most part, we won't discuss the issue of how a cognitive architecture might emerge or develop prior to this point, although, as we will see, the configuration of a complete emergent cognitive architecture isn't a straightforward task and

[8] Heinz von Foerster argues that the constituents of a cognitive architecture cannot be separated into distinct functional components: "In the stream of cognitive processes one can conceptually isolate certain components, for instance (i) the faculty to perceive, (ii) the faculty to remember, and (iii) the faculty to infer. But if one wishes to isolate these faculties functionally or locally, one is doomed to fail. Consequently, if the mechanisms that are responsible for any of these faculties are to be discovered, then the totality of cognitive processes must be considered." [122], p. 105.

[9] Ron Sun contends that "an innate structure can, but need not, be specified in an initial architecture" [117]. He argues that an innate structure does not have to be specified or involved in the computational modelling of cognition and that architectural detail may indeed result from ontogenetic development. However, he suggests that non-innate structures should be avoided as much as possible and that we should adopt a minimalist approach: an architecture should include only minimal structures and minimal learning mechanisms which should be capable of "bootstrapping all the way to a full-fledged cognitive model."

it is conceivable that epigenetic considerations might be able to shed some light on the matter.

Finally, we remind ourselves that the emergent paradigm rejects the position that cognitivism takes on two key issues: the dualism that separates the mind and body and treats them as distinct entities and the functionalism that treats cognitive mechanisms independently of the physical platform. The logical separation of mind and body, and of mechanism and physical realization, means that cognition can, in principle, be studied in isolation from the physical system in which it occurs. The emergent paradigm takes the opposite view, holding that the physical system — the body — is just as much a part of the cognitive process as are the cognitive mechanisms in the brain. Consequently, an emergent cognitive architecture will ideally reflect in some way the structure and capabilites — the morphology — of the physical body in which it is embedded and of which it is an intrinsic part. We consider these aspects in detail in Chapter 5 on embodiment.

3.2   Desirable Characteristics

When we say that an emergent cognitive architecture *ideally* reflects the form and capabilities of its associated physical body, we recognize that very few, if any, current cognitive architectures have managed to do this. There is a gap at present between what we know a cognitive architecture should be and what in fact existing architectures have managed to achieve. In this section, we focus on the ideal features of a cognitive architecture.

3.2.1   Realism

We begin with some features related to the realism of the architecture. Since a cognitivist cognitive architecture represents a unified theory of cognition, and hence a theory of human cognition, it should strive to exhibit several types of realism.[10]

First, it should enable the cognitive agent to operate in its natural environment, engaging in "everyday activities." This is referred to as *ecological realism*. It means that the architecture

[10] These different types of realism — ecological, bio-evolutionary, cognitive — as well as several other desirable characteristics of a cognitive architecture are described by Ron Sun in his paper "Desiderata for Cognitive Architectures" [117].

has to deal with many concurrent and often conflicting goals in an environment about which the agent probably doesn't know everything. In fact, that's exactly the point of cognition: being able to deal with these uncertainties and conflicts in a way that still gets the job done, whatever it is. So, ecological realism goes to the very heart of cognition.

Second, since human intelligence evolved from the capabilities of earlier primates, ideally a cognitive model of human intelligence should be reducible to a model of animal intelligence. This is *bio-evolutionary realism*. Sometimes, this is taken the other way around by focussing on simpler models of cognition as exhibited by other species — birds and rats, for example — and then attempting to scale them up to human-level cognition.

Third, a cognitive architecture should capture the essential characteristics of human cognition from several perspectives: psychology, neuroscience, and philosophy, for example. This is referred to as *cognitive realism*. To an extent, this means that the cognitive architecture, and the overall cognitive model of which it an essential part, should be complete.

Finally, as with all good science, new models should draw on, subsume, or supercede older models (this means that a cognitive architecture should strive for *inclusivity of prior perspectives*[11]).

[11] Ron Sun refers to this inclusivity as "eclecticism of methodologies and techniques" [117].

3.2.2 Behavioural Characteristics

Several behavioural and cognitive characteristics should ideally be captured by a cognitive architecture and exhibited by a cognitive system.[12] From a behavioural perspective, a cognitive architecture should not have to employ excessively complicated conceptual representations and extensive computations devoted to working through alternative strategies. The cognitive system should behave in a direct and immediate manner, making decisions and acting in an effective and timely manner. Furthermore, a cognitive system should operate one step at a time, in a sequence of actions extended over time. This gives rise to the desirable characteristic of being able to learn routine behaviours gradually, either by trial-and-error or by copying other cognitive agents.

[12] Again, these ideal behavioural and cognitive characteristics are described by Ron Sun in his paper "Desiderata for Cognitive Architectures" [117].

3.2.3 Cognitive Characteristics

As far as cognitive characteristics are concerned, a cognitive architecture should comprise two distinct types of process: one explicit, the other implicit. The explicit processes are accessible and precise whereas the implicit ones are inaccessible and imprecise. Furthermore, there should be a synergy borne of interaction between these two types of process. There are, for example, explicit and implicit learning processes and these interact.[13] To a significant extent, these cognitive characteristics reflect a hybrid approach to cognition: strict emergent approaches would not be able to deliver on the requirement for accessibility, which cognitivist approaches most certainly would. At the same time, not all cognitive architectures make use of implicit processes, such as connectionist learning, although there is an increasing trend to do so, as we will see below when we survey three current cognitive architectures.

3.2.4 Functional Capabilities

In fulfilling these roles, an ideal cognitive architecture should ideally exhibit several functional capabilities.[14]

A cognitive architecture should be able to recognize objects, situations, and events as instances of known patterns and it must be able to assign them to broader concepts or categories. It should also be able to learn new patterns and categories, modify existing ones, either by direct instruction or by experience.

Since a cognitive architecture exists to support the actions of a cognitive agent, it should provide a way to identify and represent alternative choices and then decide which are the most appropriate and select an action for execution. Ideally, it should be able to improve its decisions through learning.

It should have some perceptual capacity — vision, hearing, touch, for example[15] — and, since a cognitive agent typically has limited resources for processing information, it should have an attentive capacity to decide how to allocate these resources and to detect what is immediately relevant.

A cognitive architecture should also have some mechanism to predict situations and events, i.e. to anticipate the future. Often,

[13] These cognitive characteristics are reflected in Sun's own cognitive architecture CLARION [116, 17], in which implicit processes operate on connectionist representations and implicit processes on symbolic representations (thus, CLARION is a hybrid cognitive architectures).

[14] Pat Langley, John Laird, and Seth Rogers [120] catalogue nine functional capabilities that should be exhibited by an ideal cognitive architecture. Although they focus mainly on cognitivist cognitive architectures in their examples, the capabilities they discuss also apply for the most part to emergent systems. Ron Sun lists a similar list of twelve functional capabilities [17].

[15] There are two categories of perception: exteroception and proprioception. Exteroception includes all those modalities which sense the external world, such as vision, hearing, touch, and smell. Proprioception is concerned with sensing the status or configuration of the agent's body; whether an arm is extended and by how much, for example.

this ability will be based on an internal model of the cognitive agent's environment. Ideally, a cognitive architecture should have a mechanism to learn these models from experience and improve them over time.

To achieve goals, it must have some capability to plan actions and solve problems. A plan requires some representation of a partially-ordered sequence of actions and their effects. Incidentally, problem solving differs from planning in that it may also involve physical change in the agent's world.

The knowledge that complements a cognitive architecture constitutes the agent's beliefs about itself and its world, and planning is focussed on using this knowledge to effect some action and achieve a desired goal. The cognitive architecture should also have a reasoning mechanism which allows the cognitive system to draw inferences from these beliefs, either to maintain the beliefs or to modify them.

A cognitive architecture should have some mechanism to represent and store motor skills that can be used in the execution of an agent's actions. As always, an ideal cognitive architecture will have some way of learning these motor skills from instruction or experience.

It should be able to communicate with other agents so that they can obtain and share knowledge. This may also require a mechanism for transforming the knowledge from internal representations to a form suitable for communication.

It may also be useful for a cognitive archtitecture to have additional capabilities which are not strictly necessary but which may improve the operation of the cognitive agent. These are referred to as meta-cognition (sometimes called meta-management) functions and they are concerned with remembering (storing and recalling) the agent's cognitive experiences and reflecting on them, for example, to explain decisions, plans, or actions in terms of the cognitive steps that led to them.[16]

Finally, an ideal cognitive architecture should have some way of learning to improve the performance of all the foregoing functions and to generalize from specific experiences of the cognitive system.[17]

In summary, an ideal cognitive architecture supports at least

[16] For more details on the importance of meta-management in cognitive architectures, see Aaron Sloman's paper "Varieties of affect and the CogAff architecture schema" [16].

[17] By generalizing from specific experiences, the cognitive agent is engaging in inductive inference.

the following nine functional capabilities:

1. Recognition and categorization;

2. Decision making and choice;

3. Perception and situation assessment;

4. Prediction and monitoring;

5. Problem solving and planning;

6. Reasoning and belief maintenance;

7. Execution and action;

8. Interaction and communication;

9. Remembering, reflection, and learning.

This list is not exhaustive and one could add other functionalities: the need for multiple representations, the need for several types of memory, and the need to have different types of learning, for example. We discuss these issues in Chapters 6 and 7.

3.2.5 *Development*

One thing should strike you about the list above: it doesn't explicitly address development. That's because, for the most part, it results from research in cognitivist cognitive architectures. For emergent cognitive architecture that focus on development, Jeffrey Krichmar has identified several desirable characteristics.[18] First, he suggests that the architecture should address the dynamics of the neural element in different regions of the brain, the structure of these regions, and especially the connectivity and interaction between these regions. Second, he notes that the system should be able to effect perceptual categorization: i.e. to organize unlabelled sensory signals of all modalities into categories without prior knowledge or external instruction. In effect, this means that the system should be autonomous and, as a developmental system, it should be a model generator, rather than a model fitter.[19] Third, a developmental system should have a physical instantiation, i.e. it should be embodied, so that it is tightly coupled with its own morphology and so that it can explore its environment. Fourth, the system should engage in some behavioural task and, consequently, it should have some minimal

[18] While not specifically targetting cognitive architectures, Jeffrey Krichmar's design principles for developmental artificial brain-based devices [123, 124, 125] are directly applicable to emergent systems in general.

[19] The distinction between model generation and model fitting in cognitive systems is also emphasized by John Weng in his paper "Developmental Robotics: Theory and Experiments" [126].

set of innate behaviours or reflexes in order to explore and survive in its initial environmental niche. From this minimum set, the system can learn and adapt so that it improves its behaviour over time. Fifth, developmental systems should have a means to adapt. This implies the presence of a value system, i.e. a set of motivations that guide or govern its development.[20] These should be non-specific[21] modulatory signals that bias the dynamics of the system so that the global needs of the system are satisfied: in effect, so that the system's autonomy is preserved or enhanced.

3.2.6 Dynamics

It is clear that a cognitive system is going to be a very complex arrangement of components parts. After all, that's why an architecture is necessary in the first place. However, there is more to an architecture than just its components: there is also the manner in which they are connected with one another and the dynamic behaviour of the various components as they interact with one another and as the agent interacts with its environment. A cognitive architecture needs to be complex enough to capture these dynamics without being excessively complicated. It should incorporate only what is necessary without compromising its eco-realism. Clearly, this is a difficult balance to get right and, as we mentioned above, very few cognitive architectures fully support all of the desired characteristics at present.[22] Many challenges remain and there is a long list of issues where our understanding is inadquate. Example include[23] understanding the mechanisms for selective attention, the processes for categorization, developing support for episodic memory and processes to reflect on it, developing support for multiple knowledge representation formalisms, the inclusion of emotion in cognitive architectures to modulate cognitive behaviour, and the impact of physical embodiment on the overall cognitive process, including the agent's internal drives and goals.

[20] For an overview of the role of value systems in cognitive systems, see the paper by Kathryn Merrick "A Comparative Study of Value Systems for Self-motivated Exploration and Learning by Robots" [127] and the paper by Pierre-Yves Oudeyer, Frédéric Kaplan, and Verena Hafner "Intrinsic motivation systems for autonomous mental development" [128].

[21] Non-specific in the sense that they don't specify what actions to take.

[22] See Ron Sun's paper "The importance of cognitive architectures: an analysis based on CLARION" [17] for a more extended discussion of the degree to which contemporary cognitive architectures exhibit the desirable characteristics of an ideal architecture.

[23] This list of research challenges is taken from the paper by Pat Langley, John Laird, and Seth Rogers "Cognitive architectures: Research issues and challenges" [120].

3.3 Designing a Cognitive Architecture

Before we move on to look at some cognitive architectures that have been developed in recent years, we will first say a few words about how one might go about designing one. Given the apparent complexity of a cognitive architecture, the long list of desirable characteristics set out above, and the many research challenges we still face, it should be evident that this is not a simple matter. However, a relatively straight-forward three-step process has been proposed by Aaron Sloman and his co-workers.[24] First, the requirements of the architecture needs to be identified, partly through an analysis of several typical scenarios in which the eventual agent would demonstrate its competence. These requirements are then used to create an *architecture schema*: a task- and implementation-independent set of rules for structuring processing components and information, and controlling information flow. This schema leaves out much of the detail of the final design choices, detail which is finally filled in at the third step by an instatiation of the architecture schema in a cognitive architecture proper on the basis of a specific scenario and its attendant requirements. This process is particularly suited to cognitivist cognitive architectures because it emphasizes the logical division of task-independent processing mechanisms and structure from task-dependent knowledge.

3.4 Example Cognitive Architectures

For the remainder of the chapter, the term cognitive architecture will be used in a general sense without specific reference to the underlying paradigm, cognitivist or emergent. By this we interpret it to mean the minimal configuration of a system that is necessary for the system to exhibit cognitive capabilities and behaviours, i.e. the specification of the components in a cognitive system, their function, and their organization as a whole.

In the following, we will provide a brief overview of a sample of three cognitive architectures, one from the cognitivist paradigm of cognitive science (Soar), one from the emergent (Darwin), and one from the hybrid paradigm (ISAC).[25]

[24] The three-step process for designing a cognitive architecture is discussed in a technical report by Nick Hawes, Jeremy Wyatt, and Aaron Sloman "An architecture schema for embodied cognitive systems" [129].

[25] There are many other cognitive architectures in all three paradigms: cognitivist, emergent, and hybrid. These include for example ACT-R [109, 115], CoSy Architecture Schema [129, 130], GLAIR [131], ICARUS [132, 133] (cognitivist); Cognitive-Affective Architecture Schematic [134, 135], Global Workspace [101], iCub [136, 137], SASE [126, 138] (emergent); and CLARION [116, 17], HUMANOID [139], LIDA [140, 141], PACO-PLUS [142] (hybrid). On-line surveys of cognitive architectures can be found on the website of the Biologically Inspired Cognitive Architectures Society [143] and on the website of the University of Michigan [45]. Surveys published in the literature include an overview published by the author, Claes von Hofsten, and Luciano Fadiga in "A Survey of Artificial Cognitive Systems: Implications for the Autonomous Development of Mental Capabilities in Computational Agents" [103] and updated in *A Roadmap for Cognitive Development in Humanoid Robots* [12], and a survey by Włodzisław Duch and colleagues "Cognitive Architectures: Where do we go from here?" [144].

3.4.1   Soar

Soar[26] is a candidate Unified Theory of Cognition and, as such, it is a quintessential cognitivist cognitive architecture. It is also an iconic one, being one of the very first cognitive architectures to be developed. Futhermore, it was created by Allen Newell (the person who introduced the idea of a unified theory of cognition) and his colleagues, and has been continually enhanced over the past 25 years or so. Hence, Soar occupies a special place in the history of cognitive architectures and their continuing evolution. As we will see, the themes raised by Soar are reflected in several other cognitive architectures.

We will begin by reminding ourselves of the key ideas underpinning cognitivism. It is important to do this because Soar was built on these and the way it operates reflects the fundamental assumptions of cognitivism. We will then give a very brief sketch of the way Soar operates, just to get a feeling for the way it works.

As we have already said, in cognitivism a cognitive architecture represents the aspects of cognition that are constant over time and independent of the task. To do something, i.e. to perform a given task, a cognitivist cognitive architecture needs to be provided with the knowledge to perform the task (or it needs to acquire this knowledge for itself). This combination of a given cognitive architecture and a particular knowledge set is referred to as a *cognitive model* and it is this knowledge which populates the cognitive architecture that provides the means to perform a task or to behave in some particular way. To put it another way, cognitive behaviour equals architecture combined with content.

An architecture is a theory about what is common to the content it processes and Soar is a particular theory of what cognitive behaviours have in common. In particular, the Soar theory holds that cognitive behaviour has at least the following characteristics: it is goal-oriented, it reflects a complex environment, it requires a large amount of knowledge, and it requires the use of symbols and abstraction. The idea of abstraction is very important in cognition. It comes down to the difference between the concept of something *vs.* something in particular. For example, a shirt as

[26] For more details on the Soar cognitive architecture, please refer to the papers by Allen Newell, John Laird, and colleagues [42, 110, 111, 112, 113] and the book by John Laird [114].

a garment to provide warmth and protection *vs.* this particular blue shirt with a button-down collar and an embroidered logo on the pocket. The knowledge you have of a shirt as an abstract concept can be elicited — recalled and used — by something other than your particular perceptions in all their detail. This is referred to as a symbol (or set of symbols) and the knowledge is referred to as symbolic knowledge. The Soar cognitive architecture focusses on processing symbolic knowledge and matching it with knowledge that relate to current perceptions and actions. Let's now sketch out how it does this.

First, Soar is a production system (sometimes called a rule-based system). A production is effectively a condition-action pair and a production system is a set of production rules and a computational engine for interpreting or executing productions. Rules in Soar are called associations. Thus, the core of Soar comprises two memories, one called the long-term memory (sometimes referred to as recognition memory) which holds the productions rules, and one called working memory (also called declarative memory), which holds the attribute values that reflect Soar's perceptions and actions). In addition, there are several processes: one called *elaboration* which matches the productions and the attribute values (i.e it decides which productions can fire), one for determining the preferences to use in the decision process, and one called *chunking* which effectively learns new production rules (called *chunks*).

Soar operates in a cyclic manner with two distinct phases: a production cycle and a decision cycle. First, all productions that match the contents of declarative (working) memory fire. A production that fires may alter the state of declarative memory and cause other productions to fire. This continues until no more productions fire. At this point, the decision cycle begins and a single action is selected from several possible actions. The selection is based on stored action preferences.

Since there is no guarantee that the action preferences will be unambiguous or that they will lead to a unique action or indeed any action, the decision cycle may lead to what is known as an *impasse*. If this happens, Soar sets up a new state in a new problem space — a sub-goal — with the goal of resolving the im-

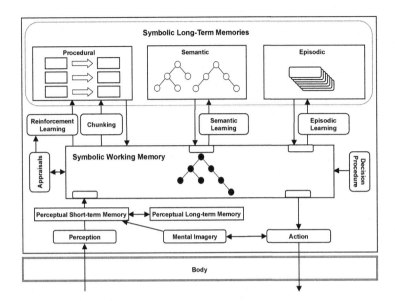

Figure 3.1: The Soar cognitive architecture, v. 9. From [114], © 2012, with permission from MIT Press.

passe. This process is known as *universal sub-goaling*. Resolving one impasse may cause others and the sub-goaling process continues. Eventually, all impasses should be resolved. In the case where the situation degenerates with Soar having insufficient knowledge to resolve the impasse, it chooses randomly between possible actions.

Whenever an impasse is resolved, Soar creates a new production rule, i.e. a new association, which summarizes the processing that occurred in the sub-state in solving the sub-goal. As we noted above, this new learned association is called a *chunk* and the Soar learning process is referred to as *chunking*.

As we said at the outset, the Soar cognitive architecture continues to evolve. While the foregoing description of Soar focussed on the production system that is so characteristic of cognitivist cognitive architectures, Soar also has the potential to be used for cognitive robotics. To facilitate this, the Soar architecture has been extended (see Figure 3.1) to embrace many of the components of emergent and hybrid cognitive architectures such as episodic memory, procedural memory, semantic memory, and associated learning techniques, e.g. reinforcement learning, as well as the crucial capability for internal simulation of perception and action using mental imagery. We discuss these topics in more detail in Chapter 7.

3.4.2   *Darwin: Neuromimetic Robotic Brain-Based Devices*

Darwin[27] is a series of robot platforms designed to experiment with developmental agents. These agents are *brain-based devices* (BBDs) which exploit a simulated nervous system that can develop spatial and episodic memory as well as recognition capabilities through autonomous experiential learning, i.e. by exploring and interacting with the world around them. BBDs are neuromimetic — they mimic the neural structure of the brain — and are closely aligned with enactive and connectionist models. However, they differ from many connectionist approaches in that they focus on the nervous system as a whole, its constituent parts, and their interaction, rather than on a neural implementation of some individual memory, control, or recognition function.

The principal neural mechanisms of a BDD are synaptic plasticity, a reward (or value) system, reentrant connectivity, dynamic synchronization of neuronal activity, and neuronal units with spatiotemporal response properties. Adaptive behaviour is achieved by the interaction of these neural mechanisms with sensorimotor correlations[28] which have been learned autonomously through active sensing and self-motion.

Different versions of Darwin exhibit different cognitive capabilities. For example, Darwin VIII is capable of discriminating reasonably simple visual targets (coloured geometric shapes) by associating them with an innately preferred auditory cue. Its simulated nervous system contains 28 neural areas, approximately 54,000 neuronal units, and approximately 1.7 million synaptic connections. The architecture comprises regions for vision (V1, V2, V4, IT), tracking (C), value or saliency (S), and audition (A). Gabor filtered images, with vertical, horizontal, and diagonal selectivity, and red-green colour filters with on-centre off-surround and off-centre on-surround receptive fields, are fed to V1. Sub-regions of V1 project topographically to V2 which in turn projects to V4. Both V2 and V4 have excitatory and in-

[27] For more details on the Darwin cognitive architecture, please refer to [123, 124, 125, 145, 146, 147].

[28] Sensorimotor correlations are sometimes referred to as *contingencies*.

hibitory reentrant connections. V4 also has a non-topographical projection back to V2 as well as a non-topographical projection to IT, which itself has reentrant adaptive connections. IT also projects non-topographically back to V4. The tracking area (C) determines the gaze direction of Darwin VIII's camera based on excitatory projections from the auditory region A. This causes Darwin to orient toward a sound source. V4 also projects to-pographically to C causing Darwin VIII to centre its gaze on a visual object. Both IT and the value system S have adaptive connections to C which facilitates the learned target selection. Adaptation is effected using the Hebbian-like learning.[29] From a behavioural perspective, Darwin VIII is conditioned to prefer one target over others by associating it with the innately preferred auditory cue and to demonstrate this preference by orienting towards the target.

[29] Specifically, the Hebbian-like learning uses the Bienenstock-Cooper-Munroe (BCM) rule [148]; also see Chapter 2, Section 2.2.1.

Darwin IX can navigate and categorize textures using arti-ficial whiskers based on a simulated neuroanatomy of the rat somatosensory system, comprising 17 areas, 1101 neuronal units, and approximately 8400 synaptic connections.

Darwin X is capable of developing spatial and episodic mem-ory based on a model of the hippocampus and surrounding regions. Its simulated nervous system contains 50 neural ar-eas, 90,000 neural units, and 1.4 million synaptic connections. It includes a visual system, head direction system, hippocam-pal formation, basal forebrain, a value/reward system based on dopaminegic function, and an action selection system. Vision is used to recognize objects and then compute their position, while odometry is used to develop head direction sensitivity.

3.4.3   ISAC

ISAC[30] — Intelligent Soft Arm Control — is a hybrid cogni-tive architecture for an upper torso humanoid robot (also called ISAC). From a software engineering perspective, ISAC is con-structed from an integrated collection of software agents and associated memories. Agents encapsulate all aspects of a com-ponent of the architecture, operate asynchonously (i.e. without a shared clock to keep the processing of all agents locked in step

[30] For a more detailed description of the ISAC cognitive architecture, please refer to "Implementation of Cognitive Control for a Humanoid Robot" by Kazuhiko Kawamura and colleagues [149].

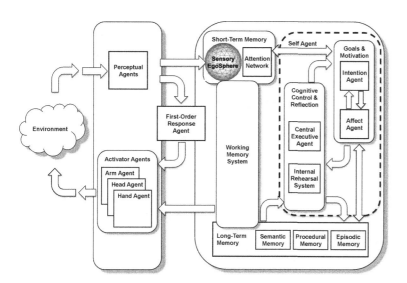

Figure 3.2: The ISAC cognitive architecture. From [149], © 2008, with permission from World Scientific Publishing Company.

with each another), and communicate with each other by passing messages.

As shown in Figure 3.2, the multi-agent ISAC cognitive architecture comprises activator agents for motion control, perceptual agents, and a First-order Response Agent (FRA) to effect reactive perception-action control. It has three memory systems: Short-term memory (STM), Long-term memory (LTM), and a working memory system (WMS).

STM has a robot-centred spatio-temporal memory of the perceptual events currently being experienced. This is called a Sensory EgoSphere (SES) and it is a discrete representation of what is happening around the robot, represented by a geodesic sphere indexed by two angles: horizontal (azimuth) and vertical (elevation). STM also has an attentional network that determines the perceptual events that are most relevant and then directs the robot's attention to them.

LTM stores information about the robot's learned skills and past experiences. LTM is made up of semantic, episodic, and procedural memory. Together, the semantic memory and episodic memory make up the robot's declarative memory of the facts it knows. On the other hand, procedural memory stores representations of the motions the robot can perform.

ISAC's episodic memory abstracts past experiences and creates links or associations between them. It has multiple layers. At the bottom, an episodic experience contains information about the external situation (i.e. task-relevant percepts from the SES), goals, emotions (in this case, internal evaluation of the perceived situation), actions, and outcomes that arise from actions, and valuations of these outcomes (e.g. how close they are to the desired goal state and any reward received at a result). Episodes are connected by links that encapsulate behaviours: transitions from one episode to another. Higher layers abstract away specific details and create links based on the transitions at lower levels. This multi-layered approach allows for efficient matching and retrieval of memories.

WMS, inspired by neuroscience models of brain function, temporarily stores information that is related to the task currently being executed. It forms a type of cache memory for STM and the information it stores, called chunks, encapsulates expectations of future reward that are learned using a neural network.

Cognitive behaviour is the responsibility of a Central Executive Agent (CEA) and an Internal Rehearsal System, a system that simulates the effects of possible actions. Together with a Goals & Motivation sub-system comprising an Intention Agent and an Affect Agent, the CEA and Internal Rehearsal System form a compound agent called the Self Agent that, along with the FRA, makes decisions and acts according to the current situation and ISAC's internal states. The CEA is responsible for cognitive control, invoking the skills required to perform some given task on the basis of the current focus of attention and past experiences. The goals are provided by the Intention Agent. Decision-making is modulated by the Affect Agent.

ISAC works the following way. Normally, the First-order Response Agent (FRA) produces reactive responses to sensory triggers. However, it is also responsible for executing tasks. When a task is assigned by a human, the FRA retrieves the skill from procedural memory in LTM that corresponds to the skill described in the task information. It then places it in the

WMS as chunks along with the current percept. The Activator Agent then executes it, suspending execution whenever a reactive response is required. If the FRA finds no matching skill for the task, the Central Executive Agent takes over, recalling from episodic memory past experiences and behaviours that contain information similar to the current task. One behaviour-percept pair is selected, based on the current percept in the SES, its relevance, and the likelihood of successful execution as determined by internal simulation in the IRS. This is then placed in working memory and the Activator Agent executes the action.

As with Soar and Darwin, there are many features in the ISAC architecture that we will discuss in greater depth later in the book, such as attention (Chapter 5, Section 5.6), the role of affect and motivation in cognition (Chapter 6, Section 6.1.1), episodic, semantic, procedural, declarative, long-term, short-term, and working memory (Chapter 7, Section 7.2), and internal simulation (Chapter 7, Section 7.4).

3.5   Cognitive Architectures — What Next?

In this chapter, we began to put some flesh on the bones of the theoretical issues set out in Chapter 2 by addressing the blueprint of every cognitive system: its architecture. This took us on a long journey, from our discussion of what a cognitive architecture means for the cognitivisit, emergent, and hybrid paradigms of cognitive science, through quite a long list of the attributes that an ideal cognitive architecture should exhibit, to short summaries of three typical cognitive architectures, one cognitivist, one emergent, and one hybrid. On the way, we've encountered many new ideas and concepts which we had to gloss over far too quickly. Our goal now is to deepen our understanding of some of these issues: autonomy, embodiment, development, learning, memory, prospection, knowledge, and representation, for example. We turn our attention first to autonomy, a concept that is difficult to model and even harder to synthesize in artificial systems.

4
Autonomy

4.1 Types of Autonomy

It is widely recognized that autonomy is a difficult concept to tie down and, like cognition, it means different things to different people.[1] To complicate matters further, there are many different ways of qualifying the concept, each suggesting a different type of autonomy. For example, you will see references to the following.

Adaptive autonomy, adjustable autonomy, agent autonomy, basic autonomy, behavioural autonomy, belief autonomy, biological autonomy, causal autonomy, constitutive autonomy, energy autonomy, mental autonomy, motivational autonomy, norm autonomy, robotic autonomy, shared autonomy, sliding autonomy, social autonomy, subservient autonomy, user autonomy, among others.

Once we have studied autonomy in more depth, we look at each of these different types of autonomy when we close the chapter in Section 4.9. In the meantime, we will pick out two of these — robotic autonomy and biological autonomy[2] — and use them as a way of organizing and explaining the other types. To get started, we need a definition[3] even if only as a basis for discussion and subsequent refinement. To that end, we will define autonomy as the degree of self-determination of a system, i.e. the degree to which a system's behaviour is not determined by the environment and, thus, the degree to which a system determines its own goals.[4] Thus, an autonomous system is not controlled by some other agent but is self-governing and self-regulating,

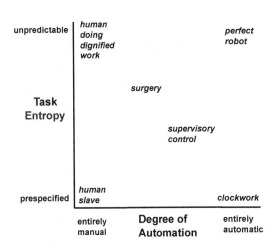

Figure 4.1: An autonomous agent — a person or a robot — can be situated in a two-dimensional space spanned in one dimension by the amount of unpredictability in the task and the working environments and in the other dimension by the degree of human assistance that is required. These two dimensions are the task entropy (i.e. uncertainty) and the degree of automation. This figure was adapted from one that appears in [156] (see Sidenote 6).

to a greater or lesser extent. Of course, it isn't a lot of use for an agent to have its own goals[5] unless it can do something about them. So, implicit in this definition is the idea that, in addition to selecting its goals, the agent can then choose how best to achieve them and that it can then act to do so.

With that preliminary definition in mind, let us now proceed to look at the two types of autonomy, robotic and biological, beginning with the former.

4.2   Robotic Autonomy

4.2.1   Strength and Degree of Autonomy

In robotics, it can be useful to categorize the capabilities of a robot on the basis of its ability to deal with uncertainty in its environment and on the extent to which a human operator assists the robot in pursuing a task and achieving some goal (see Figure 4.1).

The ability to deal with uncertainty in carrying out a task is sometimes referred to as *task entropy*.[6] At one end of the task entropy spectrum you have tasks that are completely pre-specified. There is no uncertainty at all about the task, the objects, and how to go about achieving the goal. Everything is fully known. This

[5] Willem Haselager argues that goals belong to a system "when they arise out of the on-going attempt, sustained by both the body and the control system, to maintain homeostasis ... autonomy is grounded in the formation of action patterns that result in the self-maintenance of the embodied system and it develops during embodied interaction of a system with its environment" [155]. We follow this theme at several points in the chapter; see Sidenotes 19 and 33 for more details on homeostasis and self-maintenance.

[6] The two dimensions of task entropy, i.e. environmental uncertainty, and degree of automation was suggested by Thomas Sheridan and William Verplank in 1978 in a widely-acclaimed MIT technical report dealing with human and computer control for teleoperated undersea robots [156].

is a low-entropy task. At the other end of the spectrum there is significant uncertainty about the task and there is a lot of unpredictability about what objects are present, where they are, what they look like, and what is the best way of achieving the goal of the task. This is a high-entropy task. We use the term *strength of autonomy* to denote the extent to which an autonomous system can deal with this unpredictability: strong autonomy indicates that the system can deal with considerable uncertainty in the task whereas weak autonomy indicates that it cannot.

On the other hand, the extent to which a human assists the robot reflects the degree of automation realized by the robot. We use the term *degree of autonomy*[7] to indicate the relative balance of automatic and human-assisted operation. At one end of the scale, we have entirely manual operation. This corresponds to a teleoperated robot, i.e. a robot that is controlled completely by a human operator, possibly mediated through a computer system, and typically from some distance away (hence the prefix *tele*[8]). At the other end of the scale, we have completely automatic operation, i.e. the robot operates entirely on its own, with no assistance or intervention by a human operator.

The strength of autonomy is sometimes referred to as self-sufficiency: the capability of a system to take care of itself. The degree of autonomy is sometimes referred to as self-directedness: freedom from outside control.[9] These two dimensions correspond more or less to the two dimensions of task entropy and degree of automation in Fig. 4.1 and the strength of autonomy and degree of autonomy, respectively.

Overall, we see that autonomy is a relative, relational, and situated notion: an agent is autonomous with respect to another agent, for some given action or goal, in some context, if its behaviour regarding that action or goal is not imposed by or depends on that other agent.[10]

4.2.2 *Adjustable, Shared, Sliding, and Subservient Autonomy*

Many of the types of autonomy we noted in the opening paragraph of this chapter are ways of qualifying the degree of autonomy and the relative involvement of a human with the cognitive

[7] The term *level of autonomy* is sometimes used interchangeably with *degree of autonomy* but we will just use *degree* here. The term *level of autonomy* is commonly used in the field of human–robot interaction, as noted by Michael Goodrich and Alan Schultz in their survey [157].

[8] The prefix *tele* comes from the Greek word *tēle* meaning "far off."

[9] The two dimensions of autonomy — self-sufficiency and self-directedness — were suggested by Jeffrey Bradshaw and colleagues in an article entitled "Dimensions of adjustable autonomy and mixed-initiative interaction" [158].

[10] The relational and situated nature of autonomy — the idea that an agent is autonomous *for* some action or goal and *from* something or some agent — is highlighted by Cristiano Castelfranchi in his paper "Guarantees for Autonomy in Cognitive Agent Architecture" [159] and developed by him with Rino Falcone in "Founding Autonomy: The Dialectics Between (Social) Environment and Agent's Architecture and Powers" [160] and by Cosmin Carabelea, Olivier Boissier, and Adna Florea in their article "Autonomy in Multi-agent Systems: A Classification Attempt" [161].

system in carrying out tasks and pursuing goals. For example, the terms adjustable, shared, sliding, and subservient autonomy (see Section 4.9) all suggest different degrees of autonomy in this balance of human-assisted and automatic operation in situations where the task is undertaken jointly by both humans and machines. The key point is that in these modes of autonomy the system controls its own behaviour to a greater or lesser extent but the goals are determined by the human with which it is interacting.[11]

Although the most common approach to adjustable and shared autonomy is to assign supervisory or high-level functions to the human participant and lower-level functions to the autonomous agent, there are cases where this is reversed and the human intervenes when some difficult low-level operation is needed (e.g. interpreting a visual scene[12]).

The term *sliding autonomy* is sometimes used instead of *adjustable autonomy* as it suggests the possibility of dynamically altering the level of autonomy as the circumstances require, sliding it back and forth as the task progresses. Sliding autonomy (as well as adjustable, shared, and subservient autonomy) gives rise to some interesting problems, especially when working with a team of robots. For example, the human operator may not always be aware of everything that is happening and therefore the robot may have to ask for help rather than depending on the operator to step in at just the right time when the need arises. Also, when assuming control of one of the robots, the human will take time to assess the situation. The robot needs to take account of this when making a decision to ask for help. When the human does take control of a robot, then it is important that the other robots in the team — still operating autonomously — remain autonomous and continue to work together effectively.

4.2.3   Shared Responsibility

The following ten modes of cooperation illustrate the ways in which the responsibility for carrying out a task can be shared by a human and a computer.

[11] The balance between human-assisted and automatic operation in systems that exhibit shared, subservient, adjustable, and sliding autonomy is addressed in a white paper "Measuring Performance and Intelligence of Systems with Autonomy: Metrics for Intelligence of Constructed Systems" written by Alex Meystel to explain the goals of a workshop in 2000 [162].

[12] For an example of the reversal of the usual assignment of high-level supervisory tasks to a human and low-level tasks to the autonomous agent or robot, see the "Towards perceptual shared autonomy for robotic mobile manipulation" by Benjamin Pitzer and colleagues [163]. Here, the human does the difficult job of solving low-level perceptual tasks and uses autonomous machine intelligence for the remaining high-level and low-level functions.

1. The human does the whole job up to the point of turning it over to the computer to implement.[13]
2. The computer helps by determining the options.
3. The computer helps determine options and suggests one, which the human need not follow.
4. The computer selects action and the human may or may not do it.
5. The computer selects action and implements it if the human approves.
6. The computer selects action, informs the human in plenty of time to stop it.
7. The computer does whole job and necessarily tells the human what it did.
8. The computer does whole job and tells the human what it did only if the human explicitly asks.
9. The computer does whole job and tells the human what it did and it, the computer, decides he should be told.
10. The computer does whole job if it decides it should be done, and if so tells the human, if it decides he should be told.

In these ten modes of operation, there is a continuum: from the agent — the computer or robot — being completely controlled by a human (i.e., tele-operated) all the way through to the agent being completely autonomous, operating independently of the human, not requiring any input from the human, and not requiring any approval for its actions either. This scale reflects another operational characterization of autonomy that applies in particular to robots: the tolerance of the robot to being neglected by a human operator.[14]

4.2.4 Energy Autonomy

There is one other type of autonomy that we need to mention under the heading of robotic autonomy. Quite often in robotics, and in mobile robotics in particular, when people refer to an autonomous robot they simply mean that the robot can operate for extended periods of time without being connected to an external power outlet. In other words, the robot can operate using some form of mobile power source, such as a battery or fuel cell. This is what is meant by *energy autonomy*. In this case, being autonomous has nothing to do with the unpredictability of the task

[13] This ten-level spectrum of robot autonomy was suggested by Thomas Sheridan and William Verplank in 1998, in the same technical report [156] mentioned in Sidenote 6 above. The descriptions of the ten levels are the same as they appear in Table 8.2, pp. 8-17 – 8-18, in Sheridan's and Verplank's report. Other descriptions have also been used, for example in a survey of human–robot interaction by Michael Goodrich and Alan Schultz [157] but some of the finer nuances in Sheridan's and Verplank's formulation are lost in these descriptions.

[14] The idea of *tolerance to neglect by a human operator* is suggested by Jacob Crandall *et al.* in a paper on human-robot interaction [164].

or environment or the degree of automation: it simply means energy self-sufficiency for some limited but usually lengthy period of time.

Let us move on now to consider an alternative perspective on autonomy: biological autonomy.

4.3   *Biological Autonomy*

When we consider biological — natural — autonomous entities, the issue of autonomy becomes one of survival, typically in the face of precarious conditions, i.e. environmental conditions in which the entity has to work to keep itself going as an autonomous system, both physically and organizationally as a dynamic self-sustaining entity.

Living systems face two problems: they are delicate and they are dissipative. Being delicate means that they are easily disrupted and possibly destroyed by the stronger physical forces present in their environment (including other biological agents). Consequently, living systems have to avoid these disruptions and repair or heal them when they do occur. Dissipation arises from the fact that living systems are comprised of far-from-equilibrium processes. This means that the system must have some external source of energy or matter if they are to avoid lapsing into a state of thermodynamic equilibrium. If they do succumb to this, they come to rest and cease to be able to change in response to or in anticipation of any external factors that would threaten their autonomy or their existence. Again, as with the delicacy of living systems, the dissipation inherent in far-from-equilibrium stability means that the system has to continually acquire resources, repair damage to itself, and avoid damage in the first place. All of this has to be done by the agent itself.

From this perspective, biological autonomy *is* the self-maintaining organizational characteristic of living creatures that enables them to use their own capacities to manage their interactions with the world in order to remain viable: i.e. compensate for dissipation, avoid disruption, and self-repair when necessary.[15] In other words, biological autonomy is the process by which a sys-

[15] This characterization of biological autonomy as the self-maintenant organizational characteristic of living creatures was introduced by Wayne Christensen and Cliff Hooker in their paper "An interactivist-constructivist approach to intelligence: self-directed anticipative learning" [165].

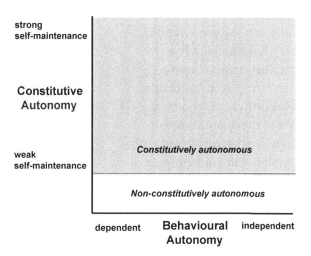

Figure 4.2: An alternative characterization of an autonomous agent situated in a two-dimensional space spanned in one dimension by behavioural autonomy and in the other by constitutive autonomy (see [151]).

tem manages — self-regulates — to maintain itself as a viable entity in the face of the precarious circumstances with which the environment continually confronts it.

4.3.1   *Behavioural and Constitutive Autonomy*

In the Section 4.2, we characterized autonomy by strength and by degree, relating these characteristics to the two-dimensional space spanned in one dimension by its robustness to the amount of unpredictability in the task and the working environment (its strength) and in the other dimension by the degree of human assistance that is required to achieve the task (its degree). These two dimensions reflect the *task entropy* (i.e. uncertainty) and the *degree of automation*, shown in Figure 4.1.

For biological autonomy, an alternative distinction can be made, differentiating between *behavioural autonomy* and *constitutive autonomy*, as illustrated in Figure 4.2 (see also Section 4.9). The behavioural dimension focusses on the degree of independence of human assistance and the extent to which the system sets its own goals, and therefore corresponds loosely to *degree of autonomy* and the dimension of *degree of automation* in Figure 4.1. The constitutive autonomy dimension focusses on the organizational characteristics that allow the system to maintain itself as an identifiable autonomous entity. Since some systems don't exhibit the requisite organizational characteristics (e.g. organizational closure; see Section 4.3.6), they aren't constitutively autonomous. These occupy the white region at the bottom of the space. Those systems that *are* constitutively autonomous can make different levels of contribution to the maintenance of their autonomy and, thus, this dimension corresponds loosely to *strength of autonomy* and the *task entropy* dimension in Figure 4.1 (in this case we interpret the term task in a very general sense to mean anything the system can be engaged in as it survives in its environment).

Behavioural autonomy focusses on the external characteristics of the system while constitutive autonomy focusses on the internal organization and the organizational processes that manage to keep the system viable and autonomous. To an extent, the external behavioural aspects mirror the degree of autonomy as it focusses on the ability of the system to function without human assistance and the system's ability to set and achieve its own goals.[16]

So, can we draw a similar parallel between constitutive autonomy and strength of autonomy as characterized by the task entropy dimension? On balance, yes we can. The concept of constitutive autonomy is concerned with maintaining the viability of the system through processes of internal organization (or self-organization). These processes can be less or more effective in dealing with the uncertainty and precariousness of the environment in which the system is embedded and in which it has to survive. Constitutive autonomy focusses on the organizational principles by which the system arises and survives as an identifiable autonomous entity in the first place and so is closely linked to the issue of autonomy in living systems, i.e. biological autonomy. In fact, the concept of constitutive autonomy derives from a very specific form of self-organization referred to as *autopoiesis* and *organizational closure*, two subjects we will discuss below in Section 4.3.6. Essentially, constitutive autonomy has more to do with the system's own internal processes than it does with the external characteristics of a precarious environment.

[16] According to Tom Froese, Nathaniel Virgo, and Eduardo Izquierdo [151] behavioural autonomy includes — in addition to the degree of independence of human assistance and the degree to which the system sets its own goals — the system's robustness and flexibility. This latter attribute reflects the way we have characterized *strength of autonomy* rather than the *degree of autonomy* and therefore it mixes degree of autonomy and strength of autonomy somewhat.

Nevertheless, the two are related: a system can't deal with uncertainty and danger if it is not organizationally — constitutively — equipped to do so.

There is just one problem that spoils this picture. The conditions for constitutive autonomy, derived from the conditions of autopoiesis and organizational closure, are very clear and strict with the result that a system is either constitutively autonomous or it isn't. From this perspective, the dimension of constitutive autonomy is less a spectrum and more of a binary classification. On the other hand, once it is constitutively autonomous, the self-organizing processes that maintain the system autonomy[17] can exhibit different levels of robustness to precarious circumstances. We will return to this issue later in the chapter when we introduce the concept of recursive self-maintenance, i.e. the contributions that a system can make to its own survival as an autonomous entity.

[17] As we will see later, it would be more accurate to refer to them as processes of self-construction and self-production.

4.3.2 Constitutive and Interactive Processes

The constitutive-behavioural distinction is sometimes cast as a difference between *constitutive* processes and *interactive* processes.[18] As we have seen, constitutive processes deal with the system itself, its organization, and its maintenance as a system through on-going processes of self-construction and self-repair. On the other hand, interactive processes deal with the interaction of the system with its environment. Both processes play complementary roles in autonomous operation of the systems. Constitutive processes are more fundamental to the autonomy of the system but both are required. Constitutive processes operate on faster time scales than interactive processes. Often robotic autonomy tends to be more concerned with interactive processes whereas biological autonomy is critically dependent on the constitutive processes. From this perspective, we can see that biological autonomy and constitutive autonomy deal very much with the same issues. For more detail and further reading, please refer to the relevant part of Section 4.9 below.

[18] The difference between *constitutive* and *interactive* processes is discussed by Tom Froese and Tom Ziemke in their paper "Enactive Artificial Intelligence: Investigating the systemic organization of life and mind" [104].

4.3.3 Homeostasis

The process of self-regulation is central to constitutive autonomy. In biological systems, the automatic regulation of physiological functions is referred to as *homeostasis*.[19] In particular, homeostatic processes regulate the operation of a system in order to keep the value of some system variables constant or within acceptable bounds. It does this by sensing any deviation from the desired value and feeding this error back to the control mechanism to correct the error. The desired value is called the *setpoint* in control theory and the use of the deviation from the desired value is called feedback. Body temperature, for example, is regulated by perspiration when we are hot (the heat used to convert the water secreted by the sweat glands into water vapour is transferred from the body to the water and this acts to lower temperature) and by shivering when we become too cold (the muscles generate heat when they twitch violently, thereby acting to increase temperature). However, homeostasis involves more than just maintaining the required system variable constant through the use of feedback as, for example, a thermostat would do when regulating the temperature of a room. A homeostatic control system itself depends on the self-regulation and will be damaged or destroyed if the regulation fails. The thermostat, on the other hand, isn't impacted at all if it isn't working correctly, although the occupants of the room it is regulating may well be.[20]

One prominent school of thought is that the autonomy of an agent is effected through a hierarchy of homeostatic self-regulatory processes, exploiting a spectrum of associated affective (i.e. emotional or feeling) states, ranging from basic reflexes linked to metabolic regulation, through drives and motives, and on to the emotions and feelings often linked to higher cognitive functions.[21] Different homeostatic processes regulate different system properties. Typically, the autonomous agent is perturbed during interactions with the world with the result that the organizational dynamics have to be adjusted. This process of adjustment is exactly what is meant by homeostasis — self-regulation — and the motives at every level of this hierarchy of homeostatic

[19] The word *homeostasis* derives from "homeo," meaning similar, and "stasis" meaning stands still or stable. It was coined by Walter Cannon in 1929 in his paper "Organization for Physiological Homeostasis" [166]. It formalizes the idea advanced in the nineteenth century by Claude Bernard that "all the vital mechanisms, however varied they may be, have only one object, that of preserving constant the conditions of life in the internal environment" [167].

[20] Homeostatic systems are more than just systems with feedback regulators. As Willem Haselager points out [155], they are self-regulating and the integrity of the homeostatic system itself depends on the self-regulation working properly. If it doesn't, the system itself will be damage or destroyed.

[21] The progression of processes of homeostasis from basic reflexes and metabolic regulation, through drives and motives, to emotions and feelings is described by Rob Lowe, Anthony Morse, and Tom Ziemke in the context of a schema for a cognitive architecture that places affect (i.e. emotion and feeling) on an equal footing with more conventional cognitive processes [134, 135]. This progression follows closely Antonio Damasio's hierarchy of levels of homeostatic regulation [168].

processes are effectively the drives that are required to return the agent to a state where its autonomy is no longer threatened. In the interaction with the world around it, the perturbations of the agent by the environment have no intrinsic value in their own right — they are just the stuff that happens to the agent as it goes about its business of survival — but for the agent this stuff, these interactions and perturbations, have a perceived value in that they act to endanger or support its autonomy. This value is conveyed though the affective aspect of these homeostatic processes and consequently the agent then attaches some value to what is an otherwise neutral world (even if it is a precarious one).

4.3.4   Allostasis

While many autonomous systems are self-governing in the sense that they adjust automatically to events in the environment and self-correct when necessary (e.g. by way of homeostasis), other autonomous systems begin to adjust *before* the event actually occurs. This form of autonomy requires a continual preparation for what might be coming next. It means that an autonomous system anticipates what events might occur in its environment and actively prepares for them so that it is capable of dealing with them if they do occur. From this perspective, autonomy requires pre-emptive action, not just reactive action, and predictive self-regulation, not just reactive self-regulation. These autonomous systems ready themselves for multiple contingencies — i.e. possible events — and have several strategies for dealing with them. They deploy them while pursing some goal or other that the system has defined for itself. This characteristic can be viewed as predictive homeostasis and is known as *allostasis*.[22] Allostasis is based on the principle that the goal of self-regulation is fitness to meet the demands placed on the autonomous system as it survives in its environment. To be fit, the system needs to be efficient: to prevent errors and minimize costs. This can be best accomplished by using prior information to anticipate the likely demands that will be placed on the system and then pre-emptively adjust all the parameters to meet this demand. Thus,

[22] The word *allostasis* was coined from the Greek roots *allo*, meaning other or departure from normal, and *stasis* meaning standing still or stable. Thus, allostasis is concerned with adapting to change in order to achieve the goal of stability in the face of uncertain circumstances. For an overview of allostasis and the relationship with homeostasis, see the articles by Peter Sterling "Principles of allostasis" [169] and "Allostasis: A model of predictive behaviour" [170]. These papers emphasize that efficient regulation requires the *anticipation* of needs and preparation to satisfy them before they arise: "The brain monitors a very large number of external and internal parameters to anticipate changing needs, evaluate priorities, and prepare the organism to satisfy them *before* they lead to errors. The brain even anticipates its own local needs, increasing flow to certain regions — before there is an error signal" [170]. Although you can view allostasis as a complementary mechanism to homeostasis, Sterling notes that it was introduced as a potential replacement for homeostasis as the core model of physiological regulation.

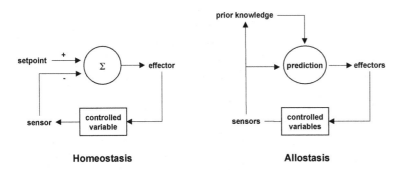

Homeostasis   **Allostasis**

Figure 4.3: *Homeostasis*
and *allostasis* are dif-
ferent approaches to
self-regulation. *Home-
ostasis* holds the value
of a controlled variable
constant by sensing its
deviation from a set-
point and using negative
feedback to correct the
error. *Allostasis* changes
the controlled variable
by predicting what level
will be needed and by
overriding local feedback
to meet the anticipated
demand. From [170],
© 2012, with permission
from Elsevier.

the goal of allostasis is efficient regulation, achieved by the brain
sensing the current state of the organism and its environment,
integrating this information with prior knowledge to change the
controlled variable by predicting what value will be needed to
meet the anticipated demand, possibly overriding local feedback to do so. These predictions are then absorbed into the prior
knowledge to improve the future predictions.

While the key difference between allostasis and homeostasis
is the predictive character of allostasis in contrast to the reactive character of homeostasis, they also differ in other important
ways. Allostatic systems adapt to change rather than resist it, as
homeostatic systems do. Also, allostasis is effected at a higher
level of organization, involving greater number of sub-systems
acting together in a coordinated manner. In contrast, mechanisms for homeostasis operate at a simpler level of negative
feedback control.[23]

The focus on predictive regulation in allostasis mirrors strongly
the anticipatory nature of cognition. We will return to this link
between autonomy and cognition in Section 4.8. Figure 4.3 illustrates the essential difference between homeostasis and allostasis.

4.3.5   Self-organization and Emergence

Autonomy is closely linked to self-organization, yet another
concept that is difficult to tie down.[24] One definition of self-
organization goes as follows.

[23] For additional details on
the way that allostasis differs
from homeostasis, see Peter
Sterling's articles "Principles
of allostasis" [169] and
"Allostasis: A model of
predictive behaviour" [170].
Refer also to the paper
"Autonomous agency, AI,
and allostasis" by Ioan
Muntean and Cory Wright
[171].

[24] Margaret Boden points
out that autonomy, self-
organization, and freedom
are three notoriously slip-
pery notions and none of
them can be properly understood without considering
the others [150]. In this
primer, we haven't considered the concept of freedom
explicitly so the reader is
encouraged to read Boden's
editorial to understand
the role freedom plays in
autonomy.

"A process in which pattern at the global level of a system emerges solely from numerous interactions among the lower-level components of the system. The rules specifying interactions among the system's components are executed using only local information, without reference to the global pattern."[25]

Typically, patterns that arise from self-organization, such as the stripes on a zebra's coat, result from the balance between processes of local activation and inhibition.

Emergence[26] also refers to a process involving interacting components in a system and the consequent generation of a global pattern. However, in this case, the global pattern emerges as something qualitatively different from the underlying assembly of components and, most significantly, is not simply a consequence of the superposition of the contributions of the individual components: they are not just "added together" to produce the result. Something else is involved in the process. It may be due to a particular form of non-linearity in the way the local components combine together to form the global pattern but it may also be due to a mutual influence between the system's local interactions and the global pattern. This form of self-organization gives rises to systems that have a clear identity or behaviour that results from two factors: (a) local-to-global determination and (b) global-to-local determination. In local-to-global determination, the emergent process has its global identity constituted and constrained by local interactions. In global-to-local determination, the global identity and its interaction with the system environment constrain the local interactions.[27] This is sometimes referred to as *emergent self-organization* (refer back to Section 1.4 and Figure 1.8). In fact, self-organization has also been defined as "the spontaneous emergence (and maintenance) of order, out of an origin that is ordered to a lesser degree."[28] This definition provides the key link between self-organization, emergence, and autonomy: that self-organization results from the intrinsic spontaneous character of the system (possibly involving interaction with the environment) rather than being imposed by some external force or agent. In other words, emergent self-organization is autonomous and, *vice versa*, autonomous systems typically involve some form of emergent self-organization.

[25] This definition of self-organization is provided in the *Encyclopedia of Cognitive Science* [85].

[26] Readers interested in a philosophical account of emergence and the related issues of anti-reductionism, subservience, and downward causation will find many insights in an article by Mark Bickhard and Donald Campbell [172]. Anil Seth provides a technical introduction to emergence [153], distinguishing between nominal, weak, and strong emergence. Strong emergence claims that macro-level properties are *in principle* not deducible from obervation of the micro-level components and they have causal powers that are irreducible, i.e. they arise only because of the existence of the emergent behaviours. These causal powers are directed at the behaviour of the components from which the emergent pattern emerges. This macro-to-micro causation is referred to as *downward causation* (see also Section 4.3.8).

[27] The concept of emergent self-organization and the two factors of local-to-global and global-to-local interaction that are involved in it are discussed in a book entitled *Enaction: Toward a New Paradigm for Cognitive Science* [48].

[28] This definition of self-organization is due to Margaret Boden and is taken from her editorial "Autonomy: What is it?" [150].

4.3.6   Self-production and Self-construction: Autopoiesis and Organizational Closure

Emergent self-organization gives rise to a special view of biological autonomy, a view that is also characterized by *self-production*. Not only is there a reciprocal local-global and global-local interaction but the nature of the interaction is to re-create the local components from which the global system arises. This is essentially constitutive autonomy. The components can be physical entities or logical organizational ones.

A system which exhibits constitutive autonomy actively generates and sustains its existence and systemic identity under precarious conditions, *i.e.* conditions which, in the absence of some appropriate form of emergent self-organization and associated behaviour, would cause the system to cease to exist and cause its identity to be destroyed.

Constitutive autonomy is closely related to a concept known as *organizational closure*. Francisco Varela famously equates organizational closure with autonomy:

> "Autonomous systems are mechanistic (dynamic) systems defined as a unity by their organization. *We shall say that autonomous systems are organizationally closed. That is, their organization is characterized by processes such that (1) the processes are related as a network, so that they recursively depend on each other in the generation and realization of the processes themselves, and (2) they constitute the system as a unity recognizable in the space (domain) in which the processes exist."*
> [97], p. 55 (emphasis in the original).

Humberto Maturana and Francisco Varela subsequently define autonomy as "the condition of subordinating all changes to the maintenance of the organization."[29]

Organizational closure is a necessary characteristic of a particular form of self-producing self-organization called *autopoiesis*,[30] that operates at the bio-chemical level, e.g., in cellular systems. Autopoiesis was introduced by Maturana and Varela, in the 1970s[31] and defined as follows.

> "*An autopoietic system is organized (defined as a unity) as a network of processes of production (transformation and destruction) of components that: (1) through their interactions and transformations continuously regenerate and realize the network of processes (relations) that produced*

[29] This definition of autonomy appears in the glossary of *Autopoiesis and Cognition — The Realization of the Living* [96].

[30] Autopoiesis, from the Greek αυτός (*autos*, meaning self) and ποιειν (*poiein*, to make or produce) and hence self-production.

[31] The seminal work of Humberto Maturana and Francisco Varela on autopoiesis is documented in *The Biology of Cognition* published in 1970 by Maturana [94] and in a subsequent paper "The Organization of the Living: a Theory of the Living Organization" in 1975 [95]. The definitive exposition is contained in jointly authored book *Autopoiesis and Cognition — The Realization of the Living* published in 1980. They also published a popular and very accessible account of their position in 1987 in a book entitled *The Tree of Knowledge — The Biological Roots of Human Understanding*. Varela's landmark book *Principles of Biological Autonomy* [97] was published in 1979.

them; and (2) constitute it (the machine) as a concrete unity in the space in which they exist by specifying the topological domain of its realization as such a network." [97], p. 13 (emphasis in the original).

Thus, autopoietic systems are quite literally self-organizing systems that self-produce. Maturana and Varela later expanded the concept to deal with autonomous systems in general and refer to it in this context as *operational closure*, rather than autopoiesis which is specific to the bio-chemical domain.[32]

4.3.7 Self-maintenance and Recursive Self-maintenance

These organizational principles are also reflected in Mark Bickhard's concepts of *self-maintenance* and *recursive self-maintenance* in far-from-equilibrium systems.[33] Self-maintenant systems contribute to the conditions which are necessary to maintain it, i.e. to keep it going. In contrast, recursive self-maintenant systems exhibit a stronger form of autonomy in that they can deploy different processes of self-maintenance depending on environmental conditions, recruiting different self-maintenant processes as conditions in the environment require. Self-maintenance and recursive self-maintenance align well with the concepts of self-organization and emergent self-organization (constitutive autonomy), respectively.

4.3.8 Continuous Reciprocal Causation

In the last three sections, there has been a recurring theme: a circular relationship between part and whole: between local factors and global factors. It appears that the characteristics of emergence and emergent self-organization are deeply dependent on dynamic re-entrant structures. This is related to the concept of continuous reciprocal causation (CRC) which occurs when some system is both continuously affecting and simultaneously being affected by activity in some other system.[34] In other words, one system causes an effect in a second system which then causes an effect in the first, reinforcing the dynamic and causing the process to continue: a very circular process. CRC can also occur in a single system. In this case, the causal contribution of

[32] The operational closure *vs.* organizational closure terminology can be confusing because in some earlier publications, e.g. [97], Varela refers to *organizational closure* but in later works (by Maturana and Varela themselves, e.g. [14], and by others, e.g. [48]) this term was subsequently replaced in favour of *operational closure*. However, as Tom Froese and Tom Ziemke note, the term operational closure is appropriate when one wants to identify any system that is identified by an observer to be self-contained and parametrically coupled with its environment but not controlled by the environment. On the other hand, organizational closure characterizes an operationally-closed system that exhibits some form of self-production or self-construction [104].

[33] The concepts of self-maintenance and recursive self-maintenance were introduced by Mark Bickhard in an article "Autonomy, Function, and Representation" [13]. Arguably, these two concepts represent a generalization of the ideas of self-construction and self-production introduced by Humberto Maturana and Francisco Varela in their processes of autopoiesis, organizational closure, and operational closure; see Sidenotes 30 and 32.

[34] The concept of *continuous reciprocal causation (CRC)* is discussed in depth in an article "Time and Mind" by Andy Clark [24].

each systemic component partially determines, and is partially determined by, the causal contributions of large numbers of other systemic components. This single-system CRC is often referred to as *circular causality* or *circular causation*.[35] While circular causality can occur between distinct sub-systems in this overall system, it more usually reflects the interaction between global system dynamics (the whole) and local system dynamics (the parts). In other words, circular causality exists *between* levels of a hierarchy of system and sub-system. This influence of macroscopic levels on microscopic levels in a system is captured in the term *downward causation*.[36] In circularly causal systems, global system behaviour influences the local behaviour of the system components and yet it is the local interaction between the components that determines the global behaviour. Thus, in biological autonomy, the degree of participation of the components of a system is determined by the global behaviour which, in turn, is determined by the interaction among the components through causal reciprocal feedback loops. Sounds confusing? It is! As you would expect, modelling circular causality and downward causation is still an open and important research problem. Despite the apparent esoteric nature of the material — and of this overview of it — it does have very practical value. To see this, we now consider a topic that is biologically inspired but is driven by the needs of software engineering.

4.4 Autonomic Systems

The ultimate goal for many people in computer technology is to produce a software-controlled system that you can simply turn on and leave to its own devices, knowing that if anything unforeseen happens, the system will sort itself out. This very desirable capability is often referred to as *autonomic computing*.[37] The originator of the term, IBM vice-president Paul Horn, defined autonomic computing systems as systems that can manage themselves, given high-level objectives from administrators. Thus, autonomic computing is strongly aligned with the concept of subservient autonomy which we discussed above. The term autonomic computing was inspired by the autonomic nervous

[35] The term *circular causality* is used by Scott Kelso in *Dynamic Patterns – The Self-Organization of Brain and Behaviour* [21] while Andy Clark uses the term *circular causation* in his book *Being There: Putting Brain, Body, and World Together Again* [23]. Clark provides an intuitive example: "Consider the way the actions of individuals in a crowd combine to initiate a rush in one direction, and the way that activity then sucks in and molds the activity of undecided individuals and maintains and reinforces the direction of collective motion."

[36] The concept of *downward causation*, i.e. that global-to-local or macroscopic-to-microscopic aspect of circular causality whereby the global system behaviour causally influences the individual system components, is discussed by Anil Seth in his article "Measuring Autonomy and Emergence via Granger Causality" [153].

[37] According to Jeffrey Kephart and David Chess in "The Vision of Autonomic Computing," an article in *IEEE Computer* [173], the term autonomic computing was introduced by IBM vice-president Paul Horn, in a keynote address to the National Academy of Engineers at Harvard University in March 2001 [174].

system found in mammals, i.e. the part of the nervous system that operates automatically to regulate the body's physiological functions such as heart-beat, breathing, and digestion. Thus, autonomic computing is also strongly aligned with self-regulatory biological autonomy, in general, and homeostasis and allostasis, in particular. Autonomic computing systems aim to exhibit several operational characteristics, including the ability to be self-configuring, self-healing, self-optimizing, and self-protecting.[38] Other characteristics have also been suggested for autonomic systems.[39] The question that arises is: how can we build such computing systems? Should we focus on autonomy or should we focus on cognition? We return to these questions in Section 4.8.

4.5 Different Scales of Autonomy

It is worth noting that autonomy applies at different scales. This means that autonomy appears at different levels in the hierarchies that are evident in natural systems. For example, think of an ant colony and an individual ant in that colony: both exhibit characteristics of autonomy but the autonomy of the ant is subservient to that of the colony. An eco-system, such as a tidal lake, may also exhibit element of autonomy over an extended period of time, self-adjusting as a complete system to keep the eco-system healthy. Within the eco-system there are many subsidiary autonomous systems: species and individuals. Again, an autonomous individual at one level of this ecological hierarchy may be subservient and therefore less autonomous when considered as a component or element of a system at a larger scale.

4.6 Goals

Looking back to the opening remark about autonomous systems setting their own goals, we note that this goal-setting capability brings with it an interesting problem. If a system, natural or artificial, is autonomous and self-controlled with its own self-determined goals, then how do you get it to do something useful for others: e.g. the people that interact with it?[40] This is particularly relevant for forms of autonomy such as adjustable, shared,

[38] The characteristics of autonomic computing — self-configuring, self-healing, self-optimizing, and self-protecting — are described in the IBM white paper "An architectural blueprint for autonomic computing" [175].

[39] James Crowley and colleagues suggest that an autonomic system should have capabilities for self-monitoring, self-regulation, self-healing, and self-description; see [176].

[40] The tension between the goals an autonomous system sets for itself and the goals another agent might wish it to pursue is addressed in a short article by the author, entitled "Reconciling Autonomy with Utility: A Roadmap and Architecture for Cognitive Development" [177].

sliding, and subservient autonomy, where the autonomy of the system is intentionally traded-off against the needs and requirements of the agents with which it interacts. This balance of goals that serve the needs of two or more agents is exemplified perfectly by symbiotic relationships where two or more autonomous systems interact to the mutual benefit of each other, while remaining focussed on their own goals and without sacrificing their autonomy.

The problem is even more acute for artificial autonomous systems because there is a clear conflict between the need for an external designer of the system and, at the same time, the need for the system to set its own goals (or take on board the goals of other agents) and autonomously maintain itself. In the words of Ezequiel Di Paolo and Hiroyuki Iizuka, "This is the apparent paradox of autonomy. The system should in some sense build itself, the designer should intervene less, but it should at the same time be more intelligently involved in setting the right processes in motion."[41]

4.7 Measuring Autonomy

So far, our treatment of autonomy has been entirely qualitative: we haven't said anything about what types of mechanisms or algorithms might be used to bring about autonomy in a system, and we haven't even hinted at some formal mathematical theory of autonomous systems. To a large extent, this is because no such theory yet exists, at least no mature proven one. However, this does not mean that people are not trying to develop one. A natural place to start in this endeavour is to try to measure autonomy since, without measurement and quantitative evaluation, it is difficult to gauge progress.[42] However, despite the importance of autonomy and the acknowledged need to be able to quantify and measure the autonomy of a system (or its degree of autonomy), such measures are very rare. One attempt to remedy this situation involves an information-theoretic[43] measure based on the observations that an autonomous system should not be governed by its environment and should determine its own goals. In particular, it uses the degree to which mutual in-

[41] This quotation is taken from Ezequiel Di Paolo's and Hiroyuki Iizuka's paper "How (not) to model autonomous behaviour" [178].

[42] The observation that it is difficult to make progress in advancing a theory of autonomy was made by Alex Meystel in a white paper explaining the goals of a workshop: "Measuring Performance and Intelligence of Systems with Autonomy: Metrics for Intelligence of Constructed Systems" [162].

[43] Information-theoretic approaches are based on a branch of mathematics known as information theory. It has its roots in the seminal work of Claude Shannon and in his famous 1948 paper "A Mathematical Theory of Communication" [179] in particular. This work builds on the formal idea of information as the reduction of uncertainty of the outcome of events and its mathematical counterpart entropy.

formation (a formal information-theoretic concept) between the system and its environment is caused by the environment or by the system itself. This is referred to as causal autonomy. [44] At present, it is not certain that this measure of autonomy can be used with autonomous systems that exhibit self-reference and that are self-maintaining. The search for such a measure remains an open problem, like so many other issues in this area. Another related quantitative measure, G-autonomy, is based on the premise that an autonomous system is not fully determined by its environment and a random system should not have a high degree of autonomy.[45] In essence, a G-autonomous system "is one for which prediction of its future evolution is enhanced by considering its own past states, as compared to predictions based on past states of a set of external variables." The G-autonomy measure is based on the statistical concept of Granger causality whereby a signal A causes a signal B if information in the history of A predicts the future of B more accurately than predictions based on the past of B alone.[46]

4.8   Autonomy and Cognition

In Chapter 1 we said that autonomy is, to a greater or lesser extent, an important attribute of cognitive systems because it allows the cognitive system to operate without human assistance. We can now see that perhaps we should have reversed the order, put it the other way around, and said that cognition may be an important attribute of an autonomous system. Certainly that is the case from the perspective of biological autonomy. This is an important switch in our understanding of cognition and autonomy so let's take the time to tease it out a little more.

Cognition, we agreed in Chapter 1, involves the ability to perceive the environment, learn from experience, anticipate the outcome of events, act to pursue goals, and adapt to changing circumstances. We also said that cognitive systems are typically autonomous, in the sense that they operate without human intervention. However, it is not necessarily the case that a cognitive system has to be autonomous: there is no reason in principle why these characteristics of perception, learning, anticipation,

[44] *Causal autonomy*: a quantitative measure of autonomy; for more details, see "Autonomy: An information theoretic perspective" by Nils Bertschinger and colleagues [154].

[45] *G-autonomy*: a quantitative measure of autonomy introduced by Anil Seth in "Measuring Autonomy and Emergence via Granger Causality" [153]. G-autonomy builds on the work of Nils Bertschinger and colleagues (see Sidenote 44).

[46] For a detailed treatment of Granger causality, see "Investigating causal relations by econometric models and cross-spectral methods" by Clive Granger [180] and "Granger causality" by Anil Seth [181].

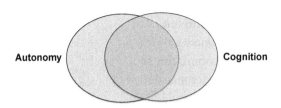

Autonomy   Cognition

Figure 4.4: Autonomy and cognition are two overlapping and complementary system attributes: some but not necessarily all cognitive systems are autonomous and, similarly, some but not all autonomous systems are cognitive.

action, and adaptation cannot be present in a system that also depends on human input. To be able to operate in such a way completely independently — autonomously — would of course be a very useful attribute, but not a necessary one.

On the other hand, an autonomous system, and certainly one that reflects biological autonomy, has a clear focus on maintaining that autonomy in a world that may not be very cooperative and helpful: in precarious circumstances, as we put it above. It may be very helpful for such an autonomous system to have some or all of the capabilities of perception, learning, anticipation (or prediction), action, and adaptation — in other words, a cognitive capability — because these will help in the process of preserving the system's autonomy. It may also be useful in pursuing the goals that the autonomous system sets for itself, implicitly or explicitly. Again, however, it is conceivable that an autonomous system could be autonomous without this cognitive capability, although in this case the autonomy may not be very robust, especially if the circumstances in which the system is operating is very precarious or changeable.

We begin to see the relationship between autonomy and cognition emerging as two overlapping and complementary system attributes: some but not necessarily all cognitive systems are autonomous and, similarly, some but not all autonomous systems are cognitive (see Figure 4.4).[47] This view fits well with the two-dimensional characterization of an autonomous agent in Figure 4.1 — task entropy *vs.* degree of automation — and the related concepts of strength of autonomy and degree of autonomy, respectively, and particularly with the idea of increased strength of autonomy being achieved through cognition.

[47] Not everyone agrees that you can have cognition without autonomy. Mark Bickhard, for example, states that "the grounds of cognition are adaptive far-from-equilibrium autonomy — recursively self-maintenant autonomy" (see Sidenote 33).

4.9   A Menagerie of Autonomies

At the beginning of this chapter, we mentioned some 19 different types of autonomy.[48] With the detailed discussion of the various issues that arise in connection with autonomy, we are now in a position to walk through these 19 types to highlight their main characteristics.

ADAPTIVE AUTONOMY The terms *adaptive autonomy* and *cognitive autonomy* are used by Tom Ziemke in his paper "On the role of emotion in biological and robotic autonomy" [152] to refer to higher levels of autonomy in complex organism. Typically, the terms refer to the processes that govern the interaction of the system with its environment, rather than the lower level constitutive processes that are concerned with the internal organization and well-being of the system (see *Constitutive Autonomy* in this section and also Section 4.3). Xabier Barandiaran uses the term *behavioural adaptive autonomy* in much the same way in a paper entitled "Behaviour Adaptive Autonomy: A Milestone on the Alife route to AI" [183]. He defines it as "homeostatic maintenance of essential variables under viability constraints [adaptivity] through a self-modulating behavioural coupling with the environment [agency], hierarchically decoupled from metabolic (constructive) processes [domain specificity]." These metabolic constructive processes are the constitutive processes referred to by Tom Ziemke.

According to this view, there are two complementary aspects of autonomy: a constructive aspect that is concerned with the constitutive metabolic processes that effectively look after the ongoing re-building of the system and an interactive aspect that is concerned with making sure the environmental conditions are right to facilitate the constructive metabolic processes. In a very loose sense, it is like the difference between eating to stay alive and foraging to make sure there is something to eat in the first place. Together, these two aspects characterize *basic autonomy*: "the organization by which far from equilibrium and thermodynamically open systems adaptively generate internal and interactive constraints to modulate the flow of matter and

[48] These are not the only types of autonomy. Others are often introduced to aid the development of a systematic treatment of the topic. For example, Michael Schillo and Klaus Fischer identify forms of autonomy based on an agent's dependency on other agents. These include skill and resource autonomy, goal autonomy, representational autonomy, deontic autonomy (subject to the authority of another agent), planning autonomy, income autonomy, exit autonomy, and processing autonomy [182]. Similarly, Cosmin Carabelea, Olivier Boissier, and Adna Florea identify five types of autonomy — user autonomy, norm autonomy, social autonomy, environment autonomy, and self autonomy — in their paper "Autonomy in Multi-agent Systems: A Classification Attempt" [161]. In "Founding Autonomy: The Dialectics Between (Social) Environment and Agent's Architecture and Powers," Cristiano Castelfranchi and Rino Falcone identify three types of autonomy depending on the level of delegation by an agent — executive autonomy, planning autonomy, and goal autonomy — as well as the concepts of norm autonomy and social autonomy.

energy required from their self-maintenance" [183] (see *Basic Autonomy* below).

ADJUSTABLE AUTONOMY The concept of adjustable autonomy is developed in an article "Dimensions of adjustable autonomy and mixed-initiative interaction" [158]. The article is based on a paper that was presented at the First International Workshop on Computational Autonomy held in July 2003 in Melbourne, Australia. The phrase is also used in the same year in the title of an article entitled "Self-organization and adjustable autonomy: Two sides of the same coin?" [184]. Adjustable autonomy is related to the concepts of shared autonomy, sliding autonomy, and subservient autonomy. All are concerned with achieving a balance of human control and agent or robot autonomy in situations where the task is undertaken jointly by both humans and machines.

AGENT AUTONOMY A complete book is devoted to the topic agent autonomy [185]. It discusses many aspects of the relationship between autonomy and agents: software entitites that independently carry out some set of functions or operations on behalf of another agent, such as a person or a computer application. The ISAC cognitive architecture we discussed in Chapter 3, Section 3.4.3, is constructed from software agents.

BASIC AUTONOMY is a term developed by Kepa Ruiz-Mirazo and Alvaro Moreno in their paper "Basic autonomy as a fundamental step in the synthesis of life" [186]. They use it to denote the capacity of a system to manage the flow of matter and energy through the system with the specific purpose of regulating — modifying and controlling — the system's own basic processes of self-construction and interaction with the environment. The process of self-construction is closely linked to the concepts of autopoiesis (see Sidenote 30) and recursive self-maintenance (see Sidenote 33), while the interactive aspect is linked to the concept of (behavioural) adaptive autonomy.

BEHAVIOURAL AUTONOMY The concept of behavioural autonomy is discussed by Tom Froese, Nathaniel Virgo, and Eduardo Izquierdo in their paper "Autonomy: a review and a reappraisal" [151]. The term is used to characterize the external aspects of autonomous systems as distinct to their internal organizational aspects (referred to as *constitutive autonomy*).

BELIEF AUTONOMY refers to the degree that an agent can exert control of its own beliefs and the degree of dependence on others to build its belief models. This complements the view of autonomy as the degree to which an agent determines its own goals and, as pointed out by Suzanne Barber and Jisun Park in their paper "Agent Belief Autonomy in Open Multi-agent Systems" [187], it highlights the mutual dependence between beliefs and goals. Barber and Park suggest that an agent should select for itself an appropriate degree of belief autonomy for a given goal on the basis of the trustworthiness of the information provided by other agents, the degree to which that information contributes to an agents's information needs, and the cost of getting that information, in terms of its timeliness.

BIOLOGICAL AUTONOMY Perhaps one of the most widely cited work on *biological autonomy* is Francisco Varela's *Principles of Biological Autonomy* [97]. It is essential reading for anyone interested in the subject. Varela's thesis is that biological autonomy arises from a specific form of self-organization referred to as *autopoiesis* (see Sidenote 30). Ezequiel Di Paulo's introduction to a special issue of *Artificial Life* — "Unbinding biological autonomy: Francisco Varela's contributions to artificial life" [188] — provides an insightful commentary on Varela's contributions. On the other hand, Tom Ziemke's "On the role of emotion in biological and robotic autonomy" [152] highlights the relationship between biological autonomy and robotic autonomy (see *Robotic Autonomy* in this section).

CAUSAL AUTONOMY The term *causal autonomy* arises in the context of quantitative measures of autonomy (see Section 4.7). In particular, the term is used by Nils Bertschinger and colleagues in their paper "Autonomy: An information theoretic

perspective" [154] to define an autonomy measure for situations
in which structure of the interaction between the autonomous
system and the environment is known. They note in their conclusions that the problem of extending their measure to cater for
the situation exhibited typically by biological autonomy, in general, and self-referential systems and self-maintaining systems,
in particular , is still open (e.g. see *Biological Autonomy* in this
section).

CONSTITUTIVE AUTONOMY The concept of *constitutive autonomy* refers to the internal organizational characteristics of an
autonomous system rather than the external behavioural aspects;
see "Autonomy: a review and a reappraisal" [151] and "Enactive
Artificial Intelligence: Investigating the systemic organization
of life and mind" [104] for more details. Constitutive autonomy
and constitutive processes are also dealt with in the main text in
Section 4.3.

ENERGY AUTONOMY (or, alternatively, *energetic autonomy*) refers
to the ability of an agent to supply its own energy needs over
an extended period of time. Chris Melhuish and colleagues, for
example, seek to achieve this in mobile robots through the use
of microbial fuel cells (MFC) that convert unrefined biomass in
the form of insect and plant material into useful energy. Their
papers "Energetically autonomous robots: Food for thought"
[189] and "Microbial fuel cells for robotics: Energy autonomy
through artificial symbiosis" [190] provide more details. More
generally, energy autonomy is set in the context of the hierarchy
of processes of homeostasis exhibited by biological autonomous
systems by Tom Ziemke and Rob Lowe in their paper "On the
Role of Emotion in Embodied Cognitive Architectures: From
Organisms to Robots" [135]. Energy autonomy occupies the
first level of the hierarchy, and the homeostatic processes act
to regulate "pre-somatic" processes, i.e. reflex-driven reactive
sensorimotor activity such as achieving a balance beween energy
consumption and energy expenditure (see also "On the role of
emotion in biological and robotic autonomy" [152]). Energy
autonomy is positioned below both motivational autonomy and

mental autonomy in the hierarchy of homeostatic processes (see the respective entries below).

MENTAL AUTONOMY refers to the processes that operate at the third level of a hierarchy on homeostatic processes (see entries on *Energy Autonomy* above and *Motivational Autonomy* below). The mental autonomy processes act to regulate the body through "somatic *simulation*," i.e. through internal simulation of behaviour and interoception (perception of internal states) using extended working memory to achieve a predictive capability that may allow higher cognitive function such as planning. Mental autonomy is positioned above motivational autonomy and energy autonomy in the hierarchy of homeostatic processes.

MOTIVATIONAL AUTONOMY refers to the processes that operate at the second level of a hierarchy on homeostatic processes (see entries on *Energy Autonomy* and *Mental Autonomy* above). The motivational autonomy processes act to regulate the body through "somatic *modulation*," i.e. through processes such as value-based learning, using basic working memory and sensorimotor association together with positive and negative rewards to achieve an elementary predictive capability. Motivational autonomy is positioned above energy autonomy and below mental autonomy in the hierarchy of homeostatic processes.

NORM AUTONOMY: a norm is a social law, convention, or organizational construct that constrains the autonomy of agents in a multi-agent system. An agent exhibits norm autonomy if it can violate or ignore that norm; see "Autonomy in Multi-agent Systems: A Classification Attempt" by Cosmin Carabelea, Olivier Boissier, and Adna Florea [161] and "Founding Autonomy: The Dialectics between (Social) Environment and Agent's Architecture and Power" by Cristiano Castelfranchi and Rino Falcone [160] . Norm autonomy is an advanced from of social autonomy (see *Social Autonomy* below).

ROBOTIC AUTONOMY The differences between *robotic autonomy* and *biological autonomy* are addressed in Tom Ziemke's paper

"On the role of emotion in biological and robotic autonomy" [152] citing a distinction made by Alvaro Moreno and colleagues between *constitutive* processes and *interactive* processes in their paper "The autonomy of biological individuals and artificial models" [191]. Tom Ziemke and Alvaro Moreno note that robotic autonomy tends to be more concerned with interactive processes whereas biological autonomy is critically dependent on the constitutive processes. Constitutive autonomy and constitutive processes are also dealt with in the main text in Section 4.3.

SHARED AUTONOMY The concepts of *shared autonomy* and *adjustable autonomy* are addressed in a paper presented by Benjamin Pitzer and colleagues at the 2011 International Conference on Robotics and Automation (ICRA) [163]. Both shared autonomy and adjustable autonomy are related to the concepts of sliding autonomy and subservient autonomy (see the respective entries below). All are concerned with achieving a balance of human control and agent or robot autonomy in situations where the task is undertaken jointly by both humans and machines.

SLIDING AUTONOMY The phrase sliding autonomy is used by Brennan Sellner and colleagues in their paper "Coordinated multi-agent teams and sliding autonomy for large-scale assembly" [192] as an alternative to *adjustable autonomy*, since it suggests the dynamic alteration of the level of autonomy with changing circumstances. Sellner suggests four different modes of operation along the sliding scale of autonomy: pure autonomy, System-Initiative Sliding Autonomy (SISA), Mixed-Initiative Sliding Autonomy (MISA), and teleoperation: "Pure autonomy does not involve the human, consisting solely of autonomous behaviors and recovery actions. In contrast, during teleoperation the human is in complete control of all aspects of all the robots. Bridging the gap between these two extremes, SISA allows the operator to intervene only when asked to do so by the autonomous system, while in MISA the human can also intervene at any time of his own volition. SISA is designed to approximate situations where the operator is a scarce resource, and must attend to multiple robot teams or other tasks. MISA, on the other hand, captures situations where humans can be more dedicated to observing the team's activities." Both sliding autonomy and adjustable autonomy are related to the concepts of shared autonomy and subservient autonomy (see respective entries in this section). Sellner's paper focusses on sliding autonomy in teams of heterogeneous robots working with the assistance of human operators.

SOCIAL AUTONOMY There are various interpretations of what social autonomy means. According to Cosmin Carabelea, Olivier Boissier, and Adna Florea in their classification of autonomy [161], *social autonomy* concerns the adoption of goals of other agents through social interaction. For social autonomy, an agent X is autonomous with respect to another agent Y for the adoption of a goal G if X can refuse the adoption of G from Y. On the other hand, Cristiano Castelfranchi and Rino Falcone define social autonomy to mean "that an agent is able and in condition to pursue and achieve a goal not depending on the other's intervention" [160]. This latter definition is similar to the concept of *user autonomy*.

SUBSERVIENT AUTONOMY The exact term used by Alex Meystel in an introduction to a workshop on measuring the performance and intelligence of autonomous systems is "subserviently autonomous" [162]. It means that while the system is capable of autonomous operation it can also be taken over and controlled by a human operator. Thus, subservient autonomy is related to the concepts of adjustable autonomy, shared autonomy, and sliding autonomy (see respective entries above). All are concerned with achieving a balance of human control and agent or robot autonomy in situations where the task is undertaken jointly by both humans and machines.

USER AUTONOMY An agent is autonomous with respect to another agent — a user — for choosing actions if it can make that choice without the user's intervention. User autonomy is one of five types of autonomy described by Cosmin Carabelea, Olivier

Boissier, and Adna Florea in their classification of autonomy [161]. The other four types are *social autonomy*, *norm autonomy*, *environment autonomy*, and *self-autonomy*. Cristiano Castelfranchi and Rino Falcone [160] refer to user autonomy as *social autonomy*.

5
Embodiment

5.1   Introduction

In the previous chapter, we explored various aspects of the relationship between autonomy and cognition. As we discovered,
this relationship poses a problem for cognitive systems. Unfortunately, there is no shortage of such problems. In Chapter 2 we
noted a number of others, one of which is the role played by the
body of an agent in cognitive activity. When a body does play a
role — and not everyone thinks it does — we often use the short-
hand *embodied cognition*.[1] In this chapter we are going to take a
closer look at embodiment.[2] We explain what is meant by embodied cognition and we discuss the ways cognition is impacted
by the physiology of a cognitive system, its evolutionary history,
its practical activity, and its social and cultural circumstances.[3]
We then consider three complementary claims that are made
about embodied cognition, referred to as the conceptualization
hypothesis, the constitution hypothesis, and the replacement
hypothesis.[4] These various aspects of embodied cognition are
founded on a two-way mutual dependence of perception and
action and we spend some time discussing the neurophysiological evidence for this dependence. We follow this with a look at
the different forms that embodiment can take. Finally, we discuss how embodied cognition is linked to the related concepts of
*situated cognition, embedded cognition, grounded cognition, extended
cognition*, and *distributed cognition*.

    To begin with, however, we remind ourselves why embod-

[1] There is a rich, varied,
and sometimes bewildering
range of writing devoted
to *embodied cognition*. The
epilogue to this chapter
gives some guidance on
where to begin.

[2] A good place to start when
reading up on *embodiment*
is Ron Chrisley's and Tom
Ziemke's article entitled,
appropriately enough, "Embodiment" [193]. It provides
a succinct synthesis, ranging over topics such as the
concept of embeddedness,
the need for representations
(or not), the importance of
dynamics, the relevance
of biology, the different
types of embodiment, the
philosophical foundations of
embodiment, and its links
with cognitive science.

[3] Michael Anderson's article
"Embodied Cognition: A
Field Guide" [194] identifies
these four aspects of em-
bodiment: physiology, evolutionary history, practical
activity, and socio-cultural
situatedness.

[4] The conceptualization, constitution, and replacement
hypotheses are discussed in
detail in Lawrence Shapiro's
book *Embodied Cognition* [83];
see Sidenotes 18, 20, and 19
in this chapter.

iment is a problematic issue in the first place. It comes down to the stance you take on cognition, i.e. whether you adhere to the cognitivist paradigm or the emergent paradigm that we discussed in Chapter 2. Let's recap each approach.

5.2   The Cognitivist Perspective on Embodiment

The essence of the cognitivist approach is that cognition comprises computational operations defined over symbolic representations and these computational operations are not tied to any given instantiation. From the perspective of the cognitivist paradigm, bodies are useful but not necessary. Cognitivist systems don't have to be embodied; they only have to be is physically instantiated and the form of this physical instantiation doesn't matter as long as it supports the computational requirements of the cognitivist cognitive architecture.[5] A physical body may facilitate exploration and learning, but it is by no means necessary.

As we noted earlier, the principled decoupling of the cognitivist computational model of cognition from its instantiation as a physical system is referred to as *computational functionalism*.[6] The chief point of computational functionalism is that the physical realization of the computational model is inconsequential to the model: any physical platform that supports the performance of the required symbolic computations will suffice, be it computer or human brain. Some caution is needed, though. Computational functionalism is the conjunction of functionalism and computationalism, both of which are quite distinct ideas.

Functionalism holds that the mind is equivalent to the functional organization of the brain. Computationalism holds that the organization of the brain is computational. Computational functionalism, then, amounts to the position that the mind is effectively the computational organization of the brain. Functionalism of itself does not entail computationalism, and neither does computationalism of itself entail functionalism. They are distinct concepts. Taken together, though, computational functionalism effectively says that the mind is the software of the brain *or any functionally equivalent system.* This is an important claim:

[5] *Cognitivist cognitive architecture*: see Chapter 3, Section 3.1.1.

[6] *Computational functionalism*: see Chapter 2, Section 2.1 and Sidenote 6.

"Computational functionalism entails that minds are multiply realizable, in the sense in which different tokens of the same type of computer program can run on different kinds of hardware. So if computational functionalism is correct, then ... mental programs can also be specified and studied independently of how they are implemented in the brain, in the same way in which one can investigate what programs are (or should be) run by digital computers without worrying about how they are physically implemented."[7]

This is not to say that cognitivist cognitive systems are not and cannot be embodied: they certainly can be and very often are. The point is that the form of the body is arbitrary, as long as it is capable of supporting the required computation. Put another way, the body of an embodied cognitivist cognitive system plays no direct part in the cognitive processes themselves: that's all down to the cognitive architecture and the knowledge comprising the cognitive model.[8]

5.3   *The Emergent Perspective on Embodiment*

The perspective from the emergent paradigm is the very opposite:[9] emergent systems are intrinsically embodied and embedded in the world around them, developing through real-time interaction with their environment. There are two complementary aspects to this development: one is the self-organization[10] of the system as a distinct entity, and the second is the coupling of that entity with its environment through interaction in the form of perception and action.[11]

Coupling, often referred to as structural coupling,[12] is a process of mutual perturbation: the cognitive system perturbing the environment and *vice versa*. It is coupling because of the mutual perturbation, and it is structural because the nature of these perturbations is determined by the physical structure of the agent and the environment. Consequently, structural coupling allows the cognitive system and its environment to adapt to each other in a mutually compatible manner. This is usually called *co-determination*.[13] The adaptation of the cognitive system over its lifetime as it gets better and better at this structural coupling is referred to as development and, more formally *ontogenetic*

[7] For an overview of the finer points of *computational functionalism*, see Gualtiero Piccinini's article "The Mind as Neural Software? Understanding Functionalism, Computationalism, and Computational Functionalism" [195]. The quotation in the main text is taken from this article.

[8] *Cognitive model*: see Chapter 3, Section 3.1.1.

[9] The striking contrast between the emergent persective and the cognitivist perspective on embodiment is evident in the *Radical Embodied Cognition Thesis* which states that the "Structured, symbolic, representational and computational view of cognition are mistaken. Embodied cognition is best studied using non-computational and non-representational ideas and explanatory schemes" [196].

[10] *Self-organization*: see Chapter 2, Section 2.2.2 and Chapter 4, Section 4.3.5.

[11] In the emergent paradigm, and in particular in the dynamical and enactive approaches, "perception, action, and cognition form a single process" [197] of self-organization in the context of environmental perturbations of the system.

[12] Alexander Riegler provides a clear explanation of *structural coupling* in his paper "When is a cognitive system embodied?", noting that it is a matter of mutual interactivity [198]. Also, see Chapter 2, Section 2.2.3 and Sidenote 56.

[13] *Co-determination*: see Chapter 2, Section 2.2.3 and Sidenote 56.

development, or just *ontogenesis*.[14] This development is effectively the (cognitive) process of establishing and enlarging the space of mutually-consistent couplings that the cognitive system can engage in, i.e. the perceptions and actions that facilitate the continued autonomy of the system.

Now, from the point of view of embodiment, here's the important part: the way the cognitive agent perceives the world — its space of possible perceptions — derives not from a predetermined, i.e. purely objective, world but from actions that the system can engage in whilst still maintaining its autonomy. In other words, it is the space of possible actions facilitated by and conditioned by the particular embodiment of the cognitive agent that determines how that cognitive agent perceives the world. Thus, through this ontogenetic development, the cognitive system constructs and develops its own understanding of the world in which it is embedded, *i.e.* its own agent-specific and body-specific knowledge of its world. This knowledge has meaning exactly because it captures the consistencies and invariances that emerge from the dynamic self-organization in the face of environmental coupling. From the emergent perspective, cognition is inseparable from bodily action because, without physical embodied action, a cognitive system cannot develop.[15] We return to this issue several more times as we continue to explore embodied cognition.

5.4   The Impact of Embodiment on Cognition

In contrasting the cognitivist and emergent stances on embodiment as we did in the previous two sections, we see they have incompatible views on the direct role played by a body in cognition. In the cognitivist paradigm, it plays no direct role. In the emergent paradigm, embodiment plays a critical role concerned with coupling the system to its world and maintaining it autonomy by enabling it to construct a meaningful understanding of the world around it and thereby act effectively in that world. This argument, though valid from the perspective of the emergent paradigm of cognitive systems, is too abstract to help us understand what embodiment means in a practical sense. So, in

[14] *Ontogenesis* and *ontogeny*: see Chapter 2, Section 2.2.3.

[15] This argument for embodiment — that cognition is inseparable from bodily action because, without physical embodied exploration, a cognitive system has no basis for development — is one of the central planks of the dynamical systems approach to development advocated by Esther Thelen and Linda Smith; for example, see Thelen's article "Time-Scale Dynamics and the Development of Embodied Cognition" [197], Thelen's and Smith's book "A Dynamic Systems Approach to the Development of Cognition and Action" [199], and their overview article "Development as a Dynamic System" [200].

this section, we will be more specific on the way embodiment might impact the operation of a cognitive system.

We set the scene by stating a general and commonly-held view: the *embodied cognition thesis*.

"Many features of cognition are embodied in that they are deeply dependent upon characteristics of the physical body of an agent, such that the agent's beyond-the-brain body plays a significant causal role, or physically constitutive role, in that agent's cognitive processing."[16]

Our aim in what follows is to look at the various strands that make up this thesis and to differentiate between the various stances people take on embodiment. In the next section, we will then formalize these stances as three distinct hypotheses on embodied cognition.

Embodiment can have an impact on cognition in a number of ways: through the physiology of the cognitive system, through its evolutionary history, through practical activity, and through its socio-cultural situation. Let's look at each of these in turn.

Proponents of embodied cognition argue that physiology, in general, and the physiology of the perceptual and motor systems of living systems, in particular, play a direct role in defining cognitive concepts and in the processes of cognitive inference. For example, whether you think something is rough or smooth depends on what you can feel through the tactile sensors in your fingertips. What looks like a steep hill to one cyclist will look like a gradual rise to another who is highly trained and has a different physiology, one that can deliver the glucose needed by the muscles at the rate they are required.

As we will see in Section 5.6, there is evidence that an agent's perceptions depend not just on what's happening in the environment and what the senses convey about these events, but they also depend on the state of the motor circuits. However, this two-way dependency between the sensory states and motor states extends further to include also a two-way dependency between perceptions (and the concepts they convey) and actions (and the goals implied by these actions). Thus, there is a blurring of the boundaries between perception, action, and cognition (at least in the emergent paradigm). These are not three

[16] The *Embodiment Thesis* is quoted from Robert Wilson's and Lucia Foglia's "Embodied Cognition" in the *Stanford Encylopedia of Philosophy* [201].

functionally-distinct processes but are rather three aspects of one global process devoted to guiding effective action and preserving the agent's autonomy. The mutual dependence of perception and action implies a dependence of cognition on the embodiment of the cognitive agent and the actions that embodiment enables. This has a far reaching consequence: *agents with different type of bodies understand the world differently.* We will return to the mutual dependence of perception and action in more detail in Section 5.6 below and also in Section 5.7 when we discuss different forms of embodiment. For now, we emphasize again that the dependence of percepts, and associated concepts constructed through cognitive activity, on the specific form of embodiment is a fundamental cornerstone of embodied cognition and emergent cognitive systems generally.

The importance of embodiment is also reflected in the *evolutionary history* of a cognitive agent. Sometimes, the way we infer things can be traced back to more primitive inference mechanisms that derive from an earlier phase in the history of evolution. Often we recruit older (in evolutionary terms) cognitive capabilities in new ways. In a sense, this is like redeploying an ultimate capability for a different new purpose.[17] What makes it possible to carry these mechanisms forward from generation to generation is the agent's embodiment, encoded in its genes.

Embodiment is also crucially important in *practical activity*. The way we go about solving problems very often relies on physical trial and error: try something out, see how well it works, and adapt (for example, by fashioning a hook to retrieve an object that is difficult to reach). The point is that these trials are dependent on your physical capabilities, the dexterity of your hands, the reach of your arms, so that cognition is bound up with the size, shape, and motor possibilities of the body as a whole. As you engage in this practical activity, you develop an understanding of the environment *in terms of your embodied action capabilities.* Thus, practical activity plays a role in giving meaning to the particular experiences of a given individual agent.

The fourth aspect of embodiment is what is referred to as *socio-cultural situatedness.* Cognitive agents are very often also social agents. This is certainly true for humans and therefore our

[17] The *ultimate* and *proximate* aspects of cognition: see Section 1.2 on the ultimate-proximate distinction.

understanding of the meaning attaching to an object or event may depend on the social and cultural context, i.e. on the way we and others have interacted in the past and the way we are expected to interact and behave now. Hand gestures, for example, can have completely different meanings in different cultures.

These four aspects — physiology, evolutionary history, practical activity, and socio-cultural situatedness — reflect the importance of embodiment in cognition. In one way or another, they all reflect the mutual dependence of perception and action and, especially, the consequent dependence of cognitive understanding on action-dependent perception.

5.5   *Three Hypotheses on Embodiment*

We can see two distinct threads in the discussion above: the impact of embodiment on the agent's understanding of the world and the direct role the body plays in the cognitive process.

The position that the physical morphology — the shape or form — and motor capabilities of a system has a direct bearing on the way the cognitive agent understands the world in which it is situated is sometimes referred to as the *conceptualization hypothesis*.[18] That is, the characteristics of an agent's body determine the concepts an organism can acquire and so agents with different type of bodies will understand the world differently.

The idea that the body (and possibly also the environment) plays a constitutive rather than a supportive role in cognitive processing, i.e. that the body is itself an integral part of cognition, is referred to as the *constitution hypothesis*.[19]

The claim made by the constitution hypothesis is stronger than that made by the conceptualization hypothesis. Cognition is not only influenced and biased by the characteristics and states of the agent's body, but the body and its dynamics augment the brain as an additional cognitive resource. In other words, the way the body is shaped and moves help it accomplish the goals of cognition without having to depend on brain-centred neural processing.

There is a third claim sometimes made by proponents of embodied cognition: that because an agent's body is engaged in

[18] In his book *Embodied Cognition* [83], Lawrence Shapiro formulates the *conceptualization hypothesis* as the claim that "the properties of an organism's body limit or constrain the concepts an organism can acquire. That is, the concepts on which an organism relies to understand its surrounding world depend on the kind of body that it has, so that were organisms to differ with respect to their bodies, they would differ as well in how they understand the world."

[19] Lawrence Shapiro formulates the *constitution hypothesis* as the claim that "the body or world plays a constitutive rather than merely causal role in cognitive processing" [83].

real-time interaction with its environment the need for representations and representational processes is removed. This is referred to as the *replacement hypothesis*.[20] The point of this hypothesis is that there is no need for the cognitive system to represent anything, computationally or otherwise, because all the information it needs is already immediately accessible as a consequence of its sensorimotor interaction. This non-representational position is neatly captured in the observation by Rodney Brooks that "the world is its own best model."[21]

The replacement hypothesis is also linked to the constitution hypothesis. If we view the body as an additional cognitive resource (i.e. the position put forward in the constitution hypothesis) it can also be argued that it is replacing some of the processing traditionally attributed to the brain. Furthermore, and this is the strong part of the replacement hypothesis, in doing this, it removes the need for brain-centred neural representations of the world with which the agent is engaging, replacing them with very different dynamical systems based perception-action couplings.[22] For example, walking gaits can be produced by an appropriately configured body without any central controller using passive dynamics,[23] coordinated activities such as flocking, herding, and hunting in groups of agents are effected again without any central control with a common shared representation, and a person can intercept and catch a ball thrown or hit high into the air simply by running with a speed and direction that induces a particular perceptual pattern as you watch the ball. Again this is done purely by matching actions with a (desired) perception and doesn't require you to to model, represent, or predict the ball's trajectory.

As well as making stronger claims than the conceptualization and constitution hypotheses, the replacement hypothesis is also more contentious, even among proponents of embodied cognition.[24] One of the arguments against the replacement hypothesis, at least in the strong form that involves the denial of the need for representations and representational processing of any kind, is that none of the examples put forward in its favour are *representation hungry*[25] in the sense that they involve problems that require the cognitive agent to act despite not having any direct

[20] The *replacement hypothesis* is usually associated with the dynamical systems approach in the emergent paradigm (see Section 2.2.2) and has strong parallels with the *Radical Embodied Cognition Thesis* (see Sidenote 9).

[21] The phrase "the world is its own best model" comes from "Elephants don't play chess" by Rodney Brooks [202].

[22] Andrew Wilson and Sabrina Golonka provide a number of arguments in favour of the replacement hypothesis in their article "Embodied cognition is not what you think it is" [203] along with several examples of how embodiment removes the need for traditional representations in cognition.

[23] For more on passive dynamic walking, see [204].

[24] The contentious nature of the *replacement hypothesis* is discussed by Margaret Wilson in her article "Six Views of Embodied Cognition" [205]. The six views are, roughly, that cognition is situated, that cognition is time-pressured, that we off-load cognitive work onto the environment, that the environment is part of the cognitive system, that cognition is for action, and that off-line cognition is body-based.

[25] The idea of *representation hungry* problems is introduced by Andy Clark's and Josefa Toribio's paper "Doing without Representing" [206] and discussed further in Andy Clark's book *Mindware* [33].

connection with the physical situation, such as is often the case in cognition, for example when anticipating the need for some action.

The conceptualization, constitution, and replacement hypotheses can also be expressed in a slightly different way, viewing the role of the body as a *constraint* on cognition, as a *distributor* for cognition, and as a *regulator* of cognitive activity.[26] The conceptualization hypothesis effectively boils down to the body constraining or conditioning cognition. The constitution hypothesis effectively says that the cognitive process is distributed between the neural and non-neural parts of the agent's physiology, either simplifying what the brain has to do or taking over responsibility for it completely. Finally, the replacement hypothesis depends on a tight coupling of cognition and action, with the body acting as a regulator of cognitive activity.

5.6   *Evidence for the Embodied Stance: The Mutual Dependence of Perception and Action*

As we noted above, two of the main contentions of embodied cognition are (a) the body of a cognitive agent is a constitutive part of the cognitive process and (b) cognitive concepts and cognitive activity therefore depend directly on the form and capabilities of the body the agent possesses. We saw that this means there is a mutual two-way dependency between sensory data and motor activity and, moreover, that the agent's perceptions and cognitive concepts depend on its actions and action capabilites. In this section, we will look at some evidence to support this position.[27]

We consider first an important aspect of cognition: visual attention. Normally, we distinguish between two types of attention: spatial attention and selective attention. Roughly speaking, these are where we direct our gaze and what sort of things are most apparent to our gaze. You would expect spatial attention to be dependent on, and only on, the what is happening in the visual field. However, it turns out that spatial attention is also dependent on what is known as oculomotor programming, i.e. the jump-like movements of the the eye — called saccades —

[26] The idea that embodiment acts as a constraint on, a distributor for, or a regulator of cognitive activity is explained in more detail in Robert Wilson's and Lucia Foglia's "Embodied Cognition" in the *Stanford Encylopedia of Philosophy* [201]. This article provides an in-depth analysis of the arguments for and against embodied cognition and a comparison with traditional computational representational, i.e. cognitivist, cognitive science.

[27] Lawrence Barsalou's article "Social Embodiment" [207] contains a rich collection of examples of dependency between the states of the body (such as posture, arm movement, facial expression) and cognitive and affective states.

as it scans the visual field and the angle of the eye in its socket. When the eye is positioned close to the limit of its rotation, and therefore cannot saccade in any further in one direction, visual attention in that direction is attenuated.[28] Selective attention, in which some objects rather than others are more apparent, is also dependent on the motor system. For example, the ability to detect an object is enhanced when the appearance and features of the object match the configuration of the agent's hands as it prepares to grasp an object.[29] This shows that an agent's current and potential actions has a direct influence on its perceptions.

The *Pinocchio Effect*[30] is another interesting example. Using a vibrator to stimulate a person's biceps while at the same time physically restraining the arm so that the person cannot move, and specifically so that he cannot not spontaneously flex as he would do naturally if left unrestrained, within a couple of seconds the person feels that their arm is more extended than it actually is. The opposite happens for triceps stimulation. Now, when you perform the experiment while the person is holding their own nose, the effect is a rather startling perception — The Pinocchio Effect — that his nose is a foot long (some people feel their fingers elongating but not their noses, and others feel both). With vibration of the triceps the perception is one of pushing their fingers through the nose into the interior of the head. This perception arises is because the artificial stimulation creates the motor (i.e. muscular) circumstances in which the arm would normally be extended. However, since the person is grasping his nose, it can't extend. So, instead, the perception the person has is that their nose is much longer than it actually is since this is the only other way of making sense of the motor state, the perceptual state, and the impression the subject has of his own body. The point here again is that our perceptions depend on the state of our embodiment; if you fool the neural system by evoking an altered embodied state, then the perceptions change accordingly.

We also have direct neurophysiological evidence of the mutual dependence of perception and action. In recent years, studies have given us a good overall view of how reaching and grasping actions are planned and executed by the monkey brain. The brain has two areas devoted to controlling the movements of

[28] The dependence of spatial attention on the angle of eye rotation is documented in a paper by Laila Craighero, Mauro Nascimben, and Luciano Fadiga: "Eye Position Affects Orienting of Visuospatial Attention" [208].

[29] The dependence of selective attention on current grasp configuration is discussed in a paper by Laila Craighero, Luciano Fadiga, and colleagues in "Movement for perception: a motor-visual attentional effect" [209]

[30] The example of the *Pinocchio Effect* shows how one can deceive someone's perceptions by providing misleading motor stimulation. The example in the text was published by James Lackner in 1988 [210] and recounted by Scott Kelso in his book *Dynamic Patterns – The Self-Organization of Brain and Behaviour* [21].

a primate: the premotor cortex and the motor cortex. The premotor cortex is the area of the brain that is active during motor planning and it influences the motor cortex which then executes the motor program comprising an action. The premotor cortex receives strong visual inputs from a region in the brain known as the inferior parietal lobule. These inputs serve a series visuomotor transformations for reaching (Area F4) and grasping (Area F5). Single neuron studies have shown that most F5 neurons code for specific goal-directed actions, rather than their constituent movements. Several F5 neurons, in addition to their motor properties, respond also to visual stimuli. These are referred to as visuomotor neurons.

The significance of this is that the premotor cortex of primates encodes actions (including implicit goals and expected states) and not just movements. In its neurophysiological sense, the term action defines a movement made in order to achieve a goal. The goal, therefore, is the fundamental property of the action rather than the specific motoric details of how it is achieved.

Now, according to their visual responses, two classes of visuomotor neurons can be distinguished within area F5: canonical neurons and mirror neurons.[31] The activity of both canonical and mirror neurons correlates with two distinct circumstances. In the case of canonical neurons, the same canonical neuron fires when a monkey sees a particular object and also when the monkey actually grasps an object with the same characteristic features. On the other hand, mirror neurons[32] are activated both when an action is performed and when the same or similar action is observed being performed by another agent. These neurons are specific to the goal of the action and not the mechanics of carrying it out.[33] So, for example, a monkey observing another monkey, or a human, reaching for a nut will cause mirror neurons in the premotor cortex to fire; these are the same neurons that fire when the monkey actually reaches for a nut. However, if the monkey observes another monkey making exactly the same movements but there is no nut present — there is no apparent goal of the reaching action — then the mirror neurons don't fire. Similarly, different motions that comprise the same goal-directed action will cause the same mirror neurons

[31] For more details on the canonical and mirror neurons in the primate brain, see "Grasping objects and grasping action meanings: the dual role of monkey rostroventral premotor cortex (area F5)" by Giacomo Rizzolatti and Luciano Fadiga [211].

[32] Mirror neurons in the monkey were discovered in the 1990s by Giacomo Rizzolatti and co-workers Luciano Fadiga, Leonardo Fogassi, and Vittorio Gallese; see, e.g., "Action Recognition in the Premotor Cortex" [212] and "Premotor cortex and the recognition of motor actions" [213]. However, because there are no equivalent invasive experiments in humans, the presence of mirror neurons in humans is inferred rather than being empirically established. For a good overview, see the review article "The Mirror-Neuron System" by Giacomo Rizzolatti and Laila Craighero [214].

[33] For more details on the goal-orientation of actions, see the article by Claes von Hoftsten: "An action perspective on motor development" [215].

to fire. It's the action that matters, not the motor activity or the movement.

Mirror neurons provide empirical neurophysiological evidence of the bi-directional interdependence action and perception — how you perceive an object depends on how you can act towards that object — and consequently how you categorize an object will depend in part on the motor capabilities and the embodiment of the agent. Thus, the presence of canonical and mirror neurons are often cited in support of the conceptualization hypothesis. However, it's not just that the shape of an object can be categorized according to how it matches, e.g., the grasp capabilities of an agent — a table-tennis ball is perceived not just as a sphere but as a *sphere-graspable-by-finger-and-thumb*, a tennis ball as a *sphere-graspable-by-a-whole-hand*, and a medicine ball as a *sphere-graspable-by-two-hands*[34] — but perhaps more importantly that the specific form of embodiment also impacts on what you can do with that object. Canonical neurons are sufficient to explain the former perception-action interdependence but mirror neurons reflect also the intention of the agent, i.e. how it anticipates it might interact with that object and, thus, they show how embodiment (they are motor neurons, after all) is directly involved in at least one of the essential characteristics of cognition, i.e. the prospective aspect. From this perspective, mirror neurons also provide some support for the constitution hypothesis.

In this context, it's important to remark on the parallel between action-dependent perception and the concept of *affordance* introduced by the ecological psychologist James J. Gibson: the idea that the perception of the potential use to which an object can be put depends on the action capabilities of the observing agent as much as it does on the object itself.[35]

While the evidence for the mutual — bi-directional — dependence of perception and action is compelling, there remains a possible concern that it doesn't make a convincing case for the embodiment of higher cognitive functions. As it happens, there is a link between the perception-action dependence and higher functions: sensorimotor processes and higher-level processes share or make use of the same neural mechanisms. Specifically, they do so when the higher-level cognitive functions engage in

[34] The table tennis and tennis ball examples of how your embodiment, and your grasp capabilities in particular, impact on the way you conceptualize and categorize objects is taken from Lawrence Shapiro's book *Embodied Cognition* [83].

[35] For a review of the link between affordance and the mirror neuron system, see the article by Serge Thill and colleagues [216]. Annemiek Barsingerhorn and colleagues summarize progress in affordance research in [217]. Also, see Chapter 2, Sidenote 52.

what is known as internal simulation (alternatively, internal emulation or rehearsal).[36] We had a brief encounter with the idea of
internal simulation earlier in Chapter 2, Section 2.2.3 and Sidenote 59, and we take it up again in this chapter in Section 5.8,
and again in more depth in Chapter 7, Section 7.5.

5.7   Types of Embodiment

We have talked quite a lot about embodiment but we haven't
said yet what exactly embodiment actually involves. What does
it mean to be embodied? What types of bodies are required?
These are the questions we address now.

We begin by making three assumptions. First, cognition involves some level of conceptual understanding of the world
with which it interacts. This is explicit in the conceptualization
hypothesis we discussed above: these concepts depend on the
nature of the agent's embodiment. Second, we assume that these
concepts are represented in some way by the cognitive system.[37]
Third, cognition involves a capacity for learning and adapting.
By constructing some model of how the world works, the cognitive agent can anticipate and act effectively. As we have noted
several times, this is a foundation of the emergent paradigm of
cognition.

Taking these three assumptions together, we see that a cognitive system must be able to construct and improve its representations. We can put this a bit stronger by saying that the
representational framework needs to fulfill two related criteria:[38]
the framework must be able to account for the possibility that
the representation is in error, and it must be able to compare the
representation with what is being represented. In other words,
the cognitive system must be able to detect its own errors. This
is necessary for error-guided behaviour and learning, two of the
principal facets of cognition, as we saw in Chapter 1. Furthermore, it must have access to its own representational contents
to allow this comparison to take place and to amend it. Some
minimal form of embodiment is required to satisfy these two
criteria. This minimal embodiment requires that the cognitive
system be capable of "full" interaction, i.e. interaction that not

[36] Read "Making Sense
of Embodied Cognition"
[218] by Henrik Svensson,
Jessica Lindblom, and Tom
Ziemke for an account of
the embodiment of higher
cognitive functions and an
explanation of the view of
cognition as body-based
simulation. The empirical
evidence presented in this
article ranges widely from
motor imagery, visual
imagery, cononical neurons,
mirror neurons, the body's
role in social interaction,
gesture, and language.

[37] The issue of *representation* is another of the many
problematic issues in cognitive systems. When we
discussed the replacement
hypothesis, we saw that
there is disagreement about
whether or not embodied
cogntive systems use internal representations of the
world around it. On the
other hand, as we discussed
Chapter 2, the nature of the
representational framework
is one of the main differences between the cognitivist
and the emergent paradigms
and there are several ways of
interpreting what is meant
by a representation. We deal
with this ambiguity in more
detail in Chapter 7.

[38] The two criteria that must
be fulfilled by a representational framework — that
it be able to account for the
possibility that it is in error
and that it be able to compare the representation with
what is being represented
— was proposed by Mark
Bickhard in his article "Is
Embodiment Necessary?"
[219]. He uses these two
criteria to argue the case for
embodied cognition.

only influences and alters the state of the world but, in turn, that these influences on the environment in turn influence the interactive process of the cognitive system. The interaction must have an impact on both the world and the agent itself. Put more formally, the agent's actions must have a causal impact on the agent's percepts of the world. This is necessary so that the agent can assess how well its representations model the world in which it is embedded. This, in turn, allows the agent to assess its anticipated actions and decide whether they are right or wrong. The assessment of anticipated action is linked to the concept of internal simulations, an issue we take up below in Section 5.8. For the moment, we stay with embodiment and, in particular, with the necessary conditions attaching to embodiment.

We have established that the minimal form of embodiment requires full interaction, with an agent's action resulting in changes in the environment that are peceptible by the agent itself. That's a good start. Can we go further? Yes, we can. At least six different types of embodiment are possible.[39] They are:

1. *Structural coupling* between agent and environment in the sense a system can be perturbed by its environment and can in turn perturb its environment;

2. *Historical embodiment* as a result of a history of structural coupling;

3. *Physical embodiment* in a structure that is capable of forcible action (this excludes software agents);

4. *Organismoid embodiment*, i.e. organism-like bodily form (e.g. humanoid robots);

5. *Organismic embodiment* of autopoietic living systems;

6. *Social embodiment* reflecting the role of the agent's body in social interaction.

These six types are increasingly more restrictive. Structural coupling entails only that the system can influence and be influenced by the physical world.[40] Historical embodiment adds the incorporation of a history of structural coupling to this level of

[39] The framework for distinguishing between different forms of embodiment — ranging from structural coupling, physical embodiment, organismoid embodiment, to organismic embodiment — was proposed by Tom Ziemke in "Disentangling Notions of Embodiment" [220]. He added historical embodiment in a subsequent paper "Are Robots Embodied?" [221], and social embodiment in a later paper "What's that thing called embodiment?" [222].

[40] The minimal form of embodiment, *structural coupling*, is taken by Kerstin Dautenhahn, Bernard Ogden, and Tom Quick [223] as the basis for an operational definition of embodiment. Their definition is significant in several ways. First, as Dautenhahn *et al.* note, it does not require embodied systems to be cognitive, conscious, intentional, made of molecules or alive, for example. Second, it provides a basis for quantifying embodiment — "regarding embodiment as a matter of degree" — based on some function of the mutual perturbation between system and environment. They suggest that there may be a measurable correlation between degree of embodiment and the extent to which cognitive behaviour is possible and, significantly, that "cognition as a phenomena arises from embodiment." Third, and somewhat paradoxically, embodiment is "freed from material constraints," leaving open the possibility of embodiment in software domains.

physical interaction so that past interactions shape the embodiment. Physical embodiment is most closely allied to conventional robot systems, with organismoid embodiment adding the constraint that the robot morphology is modelled on specific natural species or some feature of natural species. Organismic embodiment corresponds to living beings, i.e. autopoietic systems.[41] Finally, there is social embodiment, a term used to convey the role of embodiment in social interactions between agents. In this sense, it doesn't extend the continuum from structural coupling to organismic embodiment but addresses instead the mutual relationship between body states and cognitive and affective (i.e. emotional) states. This echoes the mutual dependency of perception and action which we discussed in Section 5.6 above where we mentioned in passing the dependency between the states of the body (such as posture, arm movement, facial expression) and cognitive and affective states.

There are four aspects to the relationship between body states and cognitive/affective states in social interaction. First, there is the fact that the perception of social stimuli elicits or triggers body states in the perceiving agent. Second, the perception of body states in other agents frequently evokes a tendency to mimic those states. Third, the agent's own bodily states, such as posture, constitute a powerful trigger for affective states in the agent. The three preceding aspects work together to produce a fourth aspect: the compatibility between the bodily states and the cognitive states of an agent influences the efficiency of that agent's physical and cognitive performance is several spheres, including motor control, memory, judging facial expressions, reasoning, and the effectiveness in performing tasks.[42]

The two categories of physical embodiment and organismic embodiment provide a way of contrasting two extremes of embodied cognition, one of which — *mechanistic embodiment* — is related to the Replacement Hypothesis we discussed above, and the other of which — *phenomenal embodiment* — is related to enactive systems in general, and autopoietic systems in particular.[43] Mechanistic embodiment means that cognition is embodied in the physical entity itself and, in particular in its physical control mechanism. Everything you need for cognitive activity

[41] *Autopoiesis* is an organizational characterization of what it is to be a living entity; see Chapter 4, Section 4.3.6 and Sidenote 30.

[42] As noted already, the article "Social Embodiment" by Lawrence Barsalou and co-workers [207] provides many specific and often surprising examples of the four aspects to the relationship between body states and cognitive/affective states in social interaction.

[43] These two types of embodiment — *mechanistic* and *phenomenal* — are discussed in two papers by Noel Sharkey and Tom Ziemke [224, 225] where they also refer to them as *Loebian* and *Uexküllian* embodiment, respectively, after the two biologists Jacques Loeb (1859 – 1924) and Jakob von Uexküll (1864 – 1944).

is there in the physical mechanism and there is no need to invoke mental states, representations, symbols, or the attendant need to ground these symbols in the interactive experience of the agent. The cognitive behaviour of this mechanistically embodied agent then is tied very closely to the agent's interaction with its world and, indeed, it is dependent on what happens in the environment and on what the environment does to it. The parallels with the Replacement Hypothesis are clear. On the other hand, phenomenal embodiment asserts that embodied cognition is uniquely reserved to living entities that exist in some environmental niche[44] and have a subjective or phenomenal experience of that environment. For the cognitive agent, the environment is comprehended — understood — in a strongly agent-specific manner that depends on the agent's perceptual and motoric capacities, with the agent and the environment forming a mutually-defining couple. Again, the parallel with enaction and structural coupling is clear.

The agent-specific understanding of its environment applies to embodiment in general and not just to phenomenal embodiment. If a system is embodied, it does not necessarily guarantee that cognitive concepts that result from its action-dependent perceptions will be consistent with human conceptions of cognitive behaviour. This may be quite acceptable, as long as the system does what we want it to do. However, if we want to ensure compatibility with human cognition and, specifically, compatible interaction between a human and an artificial cognitive system, then we may have to adopt the stronger version of embodiment that is consistent with the way humans are embodied: one that involves physical movement, forcible manipulation, and exploration, in human form. This is the consequence of the conceptualization hypothesis we discussed in Section 5.5 above: "to conceive of the world as a human being does requires having a body like a human being's."[45] Why? Because when two cognitive systems interact or couple, the shared understanding will only have similar meaning if the embodied experiences of the two systems are compatible.

It is important to be clear what we mean here by the term *interaction*. For the emergent stance on cognition, interaction is

[44] This subjective or phenomenal world of an agent that exhibits phenomenal embodiment is often referred to as its *Umwelt*, the term used by biologist Jakob von Uexküll [226, 227].

[45] The quotation is taken from Lawrence Shapiro's book *Embodied Cognition* [83] and encapsulates the *conceptualization hypothesis*; see Section 5.5 and Sidenote 18.

a shared activity in which the actions of each agent influence the actions of the other agents engaged in the same interaction, resulting in a mutually constructed pattern of shared behavior.[46] Such mutually constructed patterns of complementary behaviour is also emphasized in the notion of *joint action*: the coordination of individual actions by two or more people.[47] Thus, explicit meaning is not necessary for anything to be communicated in an interaction, it is simply important that the agents are mutually engaged in a sequence of actions. Meaning emerges through shared consensual experience mediated by interaction.

Finally, we note that the spectrum of embodiment from structural coupling to organismic embodiment, only the physical, organismoid, and organismic versions are sufficient to support the embodiment thesis we discussed in Section 5.4 and the conceptualization, constitution, and replacement hypotheses we discussed in Section 5.5. Historical embodiment may or may not include these three types of embodiment and structural coupling, while necessary for the other four types, doesn't have to involve body-like physicality, the cornerstone of the embodiment thesis and the three hypotheses. You can see why the topic of embodiment can be confusing and why so much has been written in the literature trying to untangle the different notions of embodiment.[48] We make our own attempt here and Section 5.10 concludes with an overview of the various meanings that are attached to embodiment and embodied cognition and various related terms, e.g. situated cognition, embedded cognition, grounded cognition, extended cognition, and distributed cognition. Before then, we consider two other inward-looking aspects of embodiment.

5.8   *Off-line Embodied Cognition*

We have already met the idea that cognition is concerned primarily with action.[49] This action-centric nature of cognition is a cornerstone of embodied cognition. However, it is clear that cognitive systems do more than just act: they think (or, equivalently, they prepare for action). This non-action cognition is sometimes referred to as *off-line cognition* or *internal simulation*. This is a thread we left off in Section 5.6 above and we take it up

[46] For more details on interaction as a shared activity, see the article by Bernard Ogden, Kerstin Dautenhahn, and Penny Stribling [228].

[47] We consider *joint action* in more detail in Chapter 9, Section 9.5.1.

[48] Tom Ziemke [229] discusses other notions of embodiment, including the distinction made by Rafael Núñez [230] between *trivial embodiment*, *material embodiment*, and *full embodiment* (respectively: cognition is directly related to biological structures, cognition is dependent on the real-time bodily interactions between an embodied agent and its environment, and the body is involved in all forms of cognition, including conceptual activity) and the distinction made by Andy Clark [231] between *simple embodiment* and *radical embodiment* (respectively: embodiment places constraints on cognitive activity, and embodiment fundamentally alters the discipline of cognitive science).

[49] The idea that cognition is action-focussed is a central tenet of embodied cognition. For example, Humberto Maturana and Francisco Varela famously stated that "cognition is effective action" [14] (see Chapter 1, Sidenote 16). Margaret Wilson puts it a slightly different way: "Cognition is for action" [205].

again here. We do so just temporarily to highlight its relevance to embodiment. We will discuss it in more detail in Chapter 7, Section 7.5.

Internal simulation is effected by preparing and executing motor programs off-line, as the brain does when anticipating events. Several possibilities can be simulated, choosing one action on the basis of an internal attention process in the light of the current flow of perceptions. Thus, the cognitive system as a whole acts as a simulator. This is the basis for the design of the HAMMER architecture, for example.[50] HAMMER accomplishes internal simulation using forward and inverse models. These encode internal sensorimotor models that the agent would utilise if it were to execute that action. The inverse model takes as input information about the current state of the system and the desired goal or goals and it outputs the control or motor commands necessary to achieve or maintain those goals. The forward model acts as a predictor. Given the current state of the system, it outputs the predicted state of the system, if the control commands provided by the inverse model were to be executed.[51] This mode of operation reflects the essential role of cognition as a multiple-timescale and multiple-outcome predictor with the express function of enabling effective action and learning effective actions. So, from this perspective, it is more appropriate to consider internal simulation as a (off-line) mode of cognition and not, for example, as a separate component of a cognitive architecture.

For embodied cognition, even off-line cognition is still body-based, i.e. the cognitive activity is still grounded in the mechanisms of sensory processing and motor control.[52] This focus on body-based internal simulation is the basis of grounded cognition, which we discuss in the section after next.

5.9 Interaction Within

There is one other twist on embodiment that we need to discuss before we close this chapter on the spectrum of ways that the environment, and the embodied agent's interaction with that environment, impacts on the cognitive capacity of that agent. So far, we have focussed on the interaction between the agent

[50] The HAMMER (Hierarchical Attentive Multiple Models for Execution and Recognition) architecture is described by Yiannis Demiris and Bassam Khadhouri in a paper of the same name [232]. A later article discusses HAMMER in the context of the mirror neuron system [233].

[51] HAMMER's internal models are actually arranged in a hierarchical manner, with models in lower levels typically working on the trajectory description level using sub-symbolic techniques and models in higher levels using symbolic techniques.

[52] With respect to off-line cognition, Michael Anderson explains that, from the perspective of embodied cognition, and despite the fact that it is decoupled from the environment in both time and space, this cognitive activity is still body-based [234]. This point is emphasized by Henrik Svensson, Jessica Lindblom, and Tom Ziemke in their article "Making Sense of Embodied Cognition: Simulation Theories of Shared Neural Mechanisms for Sensorimotor and Cognitive Processes" in which they argue that "many, if not all, higher-level cognitive processes are body-based in the sense that they make use of (partial) simulations or emulations of sensorimotor processes through the re-activation of neural circuitry that is also active in bodily perception and action" [218].

and its external environment. But bodies have insides and this raises the possibility of interaction between the agent and its own internal body, rather than interaction between agent and the external world. We touched on a similar theme when we discussed autonomic processes in Chapter 4, Section 4.4.

While some cognitive architectures restrict internal sensing — or *interoception* as it often called — to the central nervous system, other cognitive architectures extend interoception to address the affective, or emotional, aspects of the cognitive system including metabolic regulation.[53] This move reflects an increasing trend in cognitive science recognizing that cognition, as a process, is not restricted to the traditional concerns of memory, attention, and reasoning, nor yet to the co-dependence between action and perception and the approach to embodied cognition that we have been discussing so far. Instead, it also embraces the internal constitution of the embodied agent through, for example, the homeostatic processes we discussed in Chapter 4 and the affective processes that provide the internal value systems which influence the goals of autonomous cognitive agents.[54] The key point here is that a complete picture of embodiment may also have to include a place for interaction between the nervous system (and possibly also the hormone-driven endocrine system) with the body itself.[55]

5.10   *From Situated Cognition to Distributed Cognition*

One of the several cornerstones of embodied cognition is that the activities of a cognitive system — biological or artificial — take place in the context of real-world environments and involve repeated real-time interaction with that environment, in a continuous anticipatory and adaptive cycle of perception, action, and adjustment. A variety of terms are often used to convey this sense of the importance of the environment to cognition. As we noted at the beginning of the chapter, these include *situated cognition, embedded cognition, grounded cognition, extended cognition,* and *distributed cognition.* This can be confusing because, while these terms are all related, they do not necessarily mean the same thing. In this section, we will explain their differences.

[53] John Weng's SASE cognitive architecture [126, 138] focusses on the central nervous system, while the Cognitive-Affective cognitive architecture schematic developed by Rob Lowe, Anthony Morse, and Tom Ziemke [134, 135] deals explicitly with the affective and emotional aspect of cognition. Note, however, that the two perspectives are linked, especially in the case of developmental cognitive architectures, such as SASE. As we will see in Chapter 6, Section 6.1.1, development is dependent on motivation and an associated value system, and SASE builds on work that integrates a novelty-based value system with reinforcement learning [235, 236].

[54] The importance of interaction between the nervous system and the internal body is discussed by Domenico Parisi in his paper "Internal Robotics" [237]. He notes seven ways in which internal interaction differs from external interaction.

[55] Mog Stapleton calls this being "properly embodied" [238] and she argues that this requires embodied cognition to embrace interoception and a more comprehensive view of the processes that contribute an agent's cognitive capacities, including affective emotional processes.

As we saw above, one of the main arguments underpinning embodied cognition is that the cognitive agent exists in some ecological niche and that the brain-body system has evolved to take advantage of the perculiarities of its environment. Cognition, from the embodied cognition perspective and, more generally, from the emergent and enactive systems perspective, is a process that serves to support the survival of that system, to maintain its viability as an autonomous entity, in its environmental niche. Thus, we sometimes speak of the *brain-body-environment* characteristic of embodied cognition, implying that the environment is somehow a part of the cognitive process. Unfortunately, we don't always make clear to what extent the environment is involved and how it is involved. This is the nub of the problem and the source of some of the confusion. A second source of confusion is that some versions of embodiment don't require bodies! For these, realtime structural coupling is sufficient. As we mentioned above, this version of embodiment, while necessary, is not sufficient to support the embodiment thesis and the conceptualization, constitution, and replacement hypotheses we discussed in previous sections.

When we say an embodied cognitive system is *situated*,[56] we mean to draw attention to the fact that it is engaged in on-going real-time interaction with its environment: that it is structurally-coupled to the environment. Situated cognition reflects this mutual interplay between cognitive agent and environment as the system struggles to maintain its autonomy despite the precarious circumstances that the environment may present. To do this, the cognitive system sometimes exploits these circumstances.

There are several ways in which a situated embodied cognitive system makes use of the environment.[57] For example, because cognitive systems are physically-instatiated systems, they have finite capacities for, e.g., memory and attention. To overcome any problems associated with these limitations, they might off-load cognitive work onto the environment and exploit it to reduce the cognitive workload by simplifying whatever task they are engaged in. This can be something as simple as using natural landmarks to help remember your way home or something more complicated like counting with the aid of pebbles.[58]

[56] The classic text is William Clancey's book *Situated Cognition: On Human Knowledge and Computer Representations* [239]. The *Cambridge Handbook of Situated Cognition* by Philip Robbins and Murat Aydede [240] is another definitive reference.

[57] For an overview of the way that the environment is utilized in cognition, see Margaret Wilson's article "Six views of embodied cognition" [205].

[58] The use of physical aids in the environment to aid with mental processes is referred to as *symbolic off-loading*; see Andy Clark's book *Being There: Putting Brain, Body, and World Together Again* [23].

Cognitive agents also modify the environment to assist with their cognitive actions. This is referred to as *external scaffolding*.[59] We create signs to help with navigation and convey messages and we fashion and use tools to aid physical and cognitive tasks. This capacity to exploit the environment to assist cognitive activity is what is meant by *embedded cognition*. Thus, embedded cognition focusses less on the roles of the agent's body and more on the role of the physical, social, and cultural environment in cognitive activity, allowing the agent to off-load cognitive load onto elements of its environment.[60]

Grounded cognition,[61] sounds as if it might be a synonym for situated or embedded cognition but it actually means something slightly different. The focus of grounded cognition is not so much embodiment (in the sense that the body plays an active part in the cognitive process) as it is on the nature of representations that are used by a cognitive system. Grounded cognition takes issue with the symbol manipulation characteristic of the cognitivist paradigm that we discussed in Chapter 2 and especially with the assertion that these symbols, and hence the representation, are *amodal*, i.e. they are not tied directly to any particular modality such as vision, audition, or touch. Grounded cognition holds that the representations used by cognitive systems are *modal*, i.e. that they are intrinsically linked to the sensorimotor experiences of the cognitive system. From this point of view, we can see why grounded cognition is linked to embodied cognition. However, we need to be careful not to misconstrue grounded cognition as embodied cognition in the sense that we have discussed it so far. Yes, grounded cognition agrees that situated action is a key aspect of cognition and, yes, it exploits body states and sensorimotor experiences, but it does not adhere, for example, to the replacement hypothesis. On the contrary, grounded cognition does exploit symbolic representations. It is just that these representations are modal. Furthermore, a key aspect of grounded cognition is that cognition involves internal simulation, specifically modal, sensorimotor, simulation. Consequently, grounded cognition operates not just in real-time contact with the environment but it also operates off-line, independently of the body. That is not to say that situated and

[59] *External scaffolding* is discussed in Section 2.5 of Andy Clark's book [23]. Quoting an earlier book [241], he sums up the idea in his *007 Principle*: "In general, evolved creatures will neither store nor process information in costly ways when they can use the structure of the environment and their operations upon it as a convenient stand-in for the information-processing operations concerned."

[60] *Embedded cognition*, with its focus on the social and cultural environment in cognitive activity, is sometimes referred to as *social situatedness*. For more details, see the article by Jessica Lindblom and Tom Ziemke [242].

[61] For an authoritative overview of *grounded cognition*, see Lawrence Barsalou's article "Grounded Cognition" [243]. A shorter introduction can be found in his paper "Grounded cognition: Past, Present, and Future" [244]. The second article briefly reviews progress over the past 30 years and predicts some likely developments over the coming 30 years. Among these are the increasing integration of classical symbolic (cognitivist) cognitive architectures, statistical and dynamical approaches, and grounded cognition, leveraging the power of the brain's modal representational mechanisms.

embedded embodied cognitive systems don't exploit internal simulation — they can and they do — but grounded cognition makes this modal internal simulation the focus of cognition, rather than the direct involvement of the body in cognition, as encapsulated in the constitution hypothesis. Overall, we can characterize grounded cognition as agreeing with the conceptualization hypothesis, being neutral with respect to the constitution hypothesis, and positively rejecting the replacement hypothesis. In agreeing with the conceptualization hypothesis, grounded cognition also asserts that the modal representations can be based on introspective internal simulations which do not necessarily involve a faithful complete reconstruction of embodied experience. This is pivotal to the grounded cognition thesis as it allows explicitly for the inclusion of abstract concepts that are not grounded directly in specific sensorimotor experiences.

Embedded cognition involves the environment to support cognitive activity, leveraging the properties and behaviour of objects and other cognitive agents. However, there is a stronger sense in which an embodied cognitive system can involve the world around it. Instead of the environment being just something for a cognitive system to interact with, anticipate, adapt to, and even exploit to assist or amplify its cognitive activity, parts of the environment could be a direct constituent of the cognitive process itself. This view of embodied cognition is referred to as *extended cognition*.[62]

Extended cognition holds that the environment is not just the physical backdrop in which an embodied cognitive system is situated or embedded and with which it interacts, even in ways that assist with the cognitive process, but that it is a constitutent component — and an equal partner — in a bigger brain-body-environment cognitive system. Extended cognition refers not just to the extension of an agent's cognitive capability by recruiting objects or other agents in the environment but to the extension of what is the actual scope of the cognitive system itself to include the environment. *Embodied* cognition takes the critical step of asserting that cognition extends beyond the brain and includes the body. *Extended* cognition takes an additional step and claims that cognition takes place outside the body and in the larger

[62] The seminal publication on *extended cognition* is Andy Clark's and David Chalmers's article "The Extended Mind" [245]. Andy Clark has also written a book on the topic: *Supersizing the Mind: Embodiment, Action, and Cognitive Extension* [246]. A summary of this book can be found here [247]. It is also worth reading Jerry Fodor's review of the book "Where is my mind?" and Clark's response [248].

environment. For some, this is a step too far.[63]

But you can go even further. The term *distributed cognition*[64] refers to the idea that cognition takes place not only in individuals but in any system that involves interactions between people and resources in the environment. Thus, a cognitive system is characterized by dynamic self-configuration and coordination of sub-systems to accomplish some set of functions, and not by its physical extent or the range of mechanisms assumed to be responsible for cognition. So, a process is not cognitive just because it takes place in the brain, or in the brain and body, of an individual. Instead, a cognitive process is determined not by spatial location of the elements of the process but by the functional relationships among those elements. It is important that this generalization of cognition in no way precludes embodied cognitive systems as we have described them above. On the contrary, it includes them and, indeed, distributed cognition requires cognition to be embodied. But it is more than just the embodied cognition of an individual. Cognitive processes can be distributed across a group of individuals in a social group and it can involve coordinated interaction between those individuals and elements of their environment. Cognition can also be distributed through time so that events can unfold in a way that is dependent on earlier interactions. Viewed from this perspective, a social organization can be considered a cognitive architecture[65] in its own right. Furthermore, the development and operation of cognition in an individual agent depends critically on the physical embodied interaction between that agent and the material world around it, including other agents. Since social organizations operate in a historical cultural context (a context founded on interaction), distributed cognition is shaped by cultural factors, and in turn shapes them. This cultural context in which the distributed cognitive system is embedded provides an important resource for cognitive activity, akin to a corporate memory, by which the fruits of past collective experience can be shared among the individual elements — the agents — in the distributed cognitive system.

[63] The claims associated with the stance on *extended cognition* are controversial. Margaret Wilson's article "Six views of embodied cognition" [205] and Michael Anderson's "How to study the mind: an introduction to embodied cognition" provide useful critiques of these claims.

[64] The classic book on *distributed cognition* is Edwin Hutchins's 1995 text *Cognition in the Wild* [249]. An overview can be found in a paper "Distributed Cognition: Toward a New Foundation for Human-Computer Interaction Research" by James Hollan, Edwin Hutchins, and David Kirsh [250]. Hutchins article "How a Cockpit Remembers Its Speed" [251] describes an example of how a socio-technical system, specifically the cockpit of a commercial airliner comprising human pilots, instruments, and physical objects, can be viewed as a distributed cognitive system. It also contains examples of *symbolic off-loading* where certain structural features of the cockpit environment (such as the salmon speed bug) allows a conceptual cognitive task to be accomplished more simply by perceptual processes.

[65] To review what is meant by a *cognitive architecture*, refer back to Chapter 3.

A Spectrum of Embodied Cognition		
Type	**Necessary Constituents**	**Typical Characteristics**
Embodied	Depends on interpretation	Body and brain are both constitutive elements of the cognitive process
Situated	Brain	Real-time interaction with the environment
Embedded	Brain, body	Exploit the environment and other agents to assist with cognitive activities
Grounded	Brain and body	Experiential modal representations and internal simulation
Extended	Brain, body, environment	Environment is part of the cognitive system
Distributed	Brain, body, environment	Cognitive systems include environmental systems

Table 5.1: A characterization of different types of embodied cognition. Confusingly, embodied cognition is often used to refer to some or all of these different types. All involve structural coupling with the environment in some way.

5.11 Summary

So, there we have it, the many different types of embodiment — ranging from structural coupling, through historical, physical, organismoid, organismic, to social embodiment — and the various candidate aspects of embodied cognition. These are the direct involvement of the body in the cognitive process, the dependence of cognitive concepts on the specific form of the cognitive system's body, the real-time situated coupling between cognitive system and the environment, the removal the need for symbolic (or any) representations, the embedded and grounded exploitation of the environment by the cognitive system to facilitate cognitive activity and off-load cognitive work and scaffold enhanced capabilities, the interaction with the environment and especially with other cognitive systems in distributing cognitive activity, and the extension of the cognitive system to include the environment not just as a tool but as a constituent component of the cognitive process. We say 'candidate' aspects because embodiment may entail some or all of them. It depends on who is making the argument and there is no universal agreement — a recurrent theme in cognitive systems — on what embodied cognition does and does not entail.[66] However, at least the meaning of each of the various terms — situated, grounded, embedded, enactive, extended, and distributed cognition — should now be clear and Table 5.1 gives a very brief characterization of them.

We concluded Chapter 2 by remarking that, irrespective of the arguments for or against the cognitivist and emergent paradigms of cognitive systems and their long-term prospects, the current capabilities of cognitivist systems are actually more advanced. This is reflected in the state of embodied cognition which is sometimes referred to as a research program[67] rather than a mature discipline. It is a plausible and, to many, a very compelling hypothesis but, despite the fact that it is now accepted as a mainstream alternative to cognitivism, much remains to be done to establish it as an established science with well-understood engineering principles.

5.12   *Epilogue: Embodied Cognition Revisited*

We mentioned at the outset that the literature devoted to embodied cognition is rich, varied, and sometimes bewildering. Here are some ideas on where to begin reading your way into the subject.

Michael Anderson's field guide [194] is one of the definitive studies on the topic, while his article "How to study the mind: An introduction to embodied cognition" [234] provides an easy-to-read overview.

Lawrence Shapiro's book *Embodied Cognition* [83] gives a balanced and thorough explanation of the topic, touching on many of the issues covered in this chapter.

As we have seen, embodied cognition makes many claims. These are discussed very clearly in Margaret Wilson's article "Six views of embodied cognition" [205].

The influential book *The Embodied Mind* by Francisco Varela, Evan Thompson, and Eleanor Rosch [98] explains embodied cognition from the perspective of the enactive approach to emergent systems (see Chapter 2, Section 2.2.3) and is essential reading.

Paco Calvo's and Toni Gomila's *Handbook of Cognitive Science: An Embodied Approach* [252] provides a comprehensive treatment

[66] Toni Gomila's and Paco Calvo's "Directions for an embodied cognitive science: towards an integrated approach" [252] illustrates the diversity of view points on embodied cognition, giving several examples of the different taxonomies and classifications that have been suggested, many of which we have already mentioned (e.g. Margaret Wood's six views of embodied cognition [205] and Tom Ziemke's classification of embodiment [222]).

[67] Lawrence Shapiro's article "The Embodied Cognition Research Programme" [253] provides a summary of the research challenges facing embodied cognition, as well as a succinct overview of the field.

of the subject, collecting together the viewpoints of many of the most prominent researchers in the field.

Andy Clark's book *Being There: Putting Brain, Body, and World Together Again* [23] is one of the first books to offer a complete synthesis of the arguments for embodied, embedded, activity-focussed view of cognition, a view that eliminates any hard boundaries between perception, cognition, and action and highlights their interdependence.

A special issue of *Cognitive Systems Research* provides a snapshot of the research challenges in situated and embodied cognition [254]. Even though it was published in 2002, all of the issues raised are still live today.

Finally, there is a comprehensive tutorial on embodied cognition, including video lectures, on the website of the European Network for the Advancement of Artificial Cognitive Systems, Interaction, and Robotics [255].

6

Development and Learning

6.1   Development

We saw in Chapter 2 that development is a key aspect of the emergent paradigm of cognitive systems[1] and we are all aware that humans develop as they grow, especially during infancy. The degree of development is often surprising. Table 6.1 picks out just some of the milestones in the development of an infant,[2] focussing on the capabilities that are necessary for a child to acquire the ability to notice that someone might need help. This itself is just a part of the developmental process by which a child acquires an ability to actively help others and eventually collaborate with them (an aspect of cognition to which we alluded in Chapter 1; see Figure 1.5). Development is an increasingly important aspect of artificial cognitive systems, in general, and robotic systems, in particular.[3]

   Given that development is an important part of cognition, what drives the development process and what factors motivate development? Put more formally, how do innate mechanisms enable development so that ever-richer cognitive capabilities emerge, consolidate, and combine to produce an embodied agent capable of flexible anticipatory interaction? What type of phylogeny (or cognitive architecture) is needed to facilitate development? We look to developmental psychology to provide some guidance on the necessary phyology, the process of ontogenetic development, the balance between the two, and the factors that drive development.[4] First, let's discuss the concept of

[1] We remarked in Chapter 1, Sidenote 22, that, reflecting the importance of learning and development, both have been included in a revision of Marr's three-level hierarchy of understanding (above the computational theory level — the original top level — and below a new top level: evolution) [19].

[2] Infants are often referred to as *neonates* in the developmental psychology literature.

[3] For overviews of the developmental approach in robotics, see the surveys by Max Lungarella and colleagues [276] and Minoru Asada and colleagues [277].

[4] Read "Human Sensori-Motor Development and Artificial Systems" by Giulio Sandini and colleagues [278] for a good overview of how an understanding of human development inspires artificial development in robots.

Newborns

Newborns gaze longer when a person looks directly at them [256].

Newborns are attracted to people (i.e. face and voice) [257].

Newborns prefer biological motion [258].

Newborns preferentially orient toward faces [259, 260].

Newborns prefer human voices to other sounds [261].

Early Development

$2\frac{1}{2}$ months: infants can discriminate a familiar adult's expressions if presented with multimodal expressions [262].

3 months: infants engage mutual gaze with adults, i.e. both agents attend to each other's eyes simultaneously [263].

3–4 months: infants have the ability to discriminate among a few photographed, static facial expressions [264].

4 months: infants presented with multimodal expressions can discriminate some adult's expressions [265, 266].

5 months: infants discriminate auditory-only displays of affect [266].

6 months: infants can perceive approximate direction of attention of others (i.e. to the left or to the right) [267].

10–12 months: infants show the first strong evidence of understanding the feelings of others.

9 months: infants can accurately detect the direction of the adult's gaze [263].

12 months: infants look at the object fixated by the adults [268].

12 months: infants consider eye rather than head direction [269].

12 months: Children start to understand pointing as an object-directed action [270].

12 months: Children anticipate with gaze the goal of a feeding action [271].

Later Development

18 months: children start to follow an adult's gaze outside their own field of view [263].

18 months: children perceive from emotions that a person wants something [272].

18 months: infants can infer what another person is trying to achieve (even if the attempt is unsuccessful) [273, 274].

18 months: infants altruistically (*instrumentally*) help adults when they are having problems achieving a goal [275].

Table 6.1: Selected milestones in the development of a human infant, highlighting some of those that are involved in interaction and especially those that lead to an ability to collaborate with others. The innate skills in newborns provide a sensitivity to characteristics of the external world that maximize an infant's chances of interacting with others [257]. This material was compiled by Alessandra Sciutti, Istituto Italiano di Tecnologia (IIT).

development a little more.

Development arises due to changes in the central nervous system as a result of dynamic interaction with the environment. Development results in new forms of action and the acquisition of predictive control of these actions. The predictive part is important: to control actions effectively, a cognitive agent needs to be able to anticipate the outcome of these actions, as well as anticipating the need for them in the first place. It seems that cognition is the way an agent manages to align these two forward-looking views of its interaction with the environment in which it is embedded. To act effectively, an agent must be

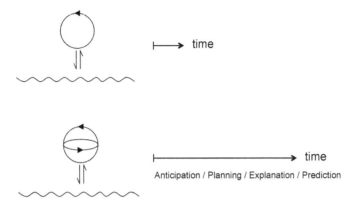

time

time

Anticipation / Planning / Explanation / Prediction

Figure 6.1: The time horizon of a cognitive system increases with development. The top half of the figure depicts a system without a central nervous system (see Chapter 2, Figure 2.3); it has little or no prospective capacity. The bottom half depicts a system with a central nervous system, and hence one capable of development (see Chapter 2, Figure 2.4); it has considerably greater prospective capacity.

able to infer upcoming events. We call this *prospection*. For example, when a cognitive agent, and specifically a human agent, practices some new action repetitively, the focus isn't as you might think on establishing fixed patterns of movement — muscle memory as it is sometimes called — but on establishing the scope for prospective control of these movements in the context of the goals of an action.[5] Thus, sensorimotor skills that are learned early on are gradually developed. The progressive development of innate skills or skills that are learned early on is sometimes referred to as *scaffolding*. This development, particularly the prospective aspect, is accelerated by internal simulation, i.e. mentally rehearsing — consciously or subconsciously — the execution of actions and inferring the likely outcome of those actions. Thus, cognition itself can be viewed as a developmental process through which the system becomes progressively more skilled and acquires the ability to understand events, contexts, and actions, initially dealing with immediate situations and increasingly acquiring a predictive, i.e. prospective, capability (see Figure 6.1).

Like cognition, development is a system-wide process: the development of action and perception, the development of the nervous system, and the development and growth of the body all mutually influence each other as increasingly-sophisticated and increasingly prospective (i.e. future-oriented) action capabilities are acquired.[6]

[5] The importance of prospection (i.e. anticipation) in motor control, especially when viewed in the context of carrying out goal-directed actions, is developed by Ed Reed in his book *Encountering the world: Toward an ecological psychology* [279]; also see the review by Gavan Lintern [280].

[6] Claes von Hofsten's paper "Action in Development" provides an accessible overview of the importance of prospective goal-directed action in cognitive development and the role of exploratory and social motivations in the developmental process [281]. A more in-depth treatment can be found in his article "Action in Infancy: A Foundation for Cognitive Development" [282].

6.1.1 *Motivation*

The development of an autonomous agent is crucially dependent on motives. They determine the goals of actions and provide the drive for achieving them. As we noted in Chapter 3, Section 3.2.5, motivations imply the presence of a value system that guide or govern development.[7] The two most important motives that drive actions, and thus development, are social and exploratory. They both function from birth and provide the driving force for action throughout life.

The social motive puts the subject in a broader context of other humans that provide comfort, security, and satisfaction, from which the subject can learn new skills, find out new things about the world, and exchange information through communication. The social motive is so important that it has been suggested that without it a person will stop developing altogether. The social motive is expressed from birth in the tendency to fixate social stimuli, imitate basic gestures, and engage in social interaction. Social motives include the need to belong, self-preservation, and cognitive consistency with others.[8]

There are at least two exploratory motives. The first one has to do with finding out about the surrounding world: the discovery of novelty and regularity in the world. The second exploratory motive has to do with finding out about the potential of one's own action capabilities.

New and interesting objects and events attract infants' visual attention, but after a few exposures they are not attracted any more. Infants also have a strong motivation to discover what they can do with objects, especially with respect to their own sensorimotor capabilities and the particular characteristics of their embodiment. Put a little more formally, infants have a strong motivation to discover the affordances of objects in their surroundings in the context of new actions they can perform with them.[9]

The motivation to seek new ways of doing things is so strong that it often overrides ways achieving a goal that have already become established through previous development. For example, infants persist in trying to walk even when they can get about

[7] For an overview of the role of value systems in artificial cognitive systems, see the paper "A Comparative Study of Value Systems for Self-motivated Exploration and Learning by Robots" by Kathryn Merrick [127] and "Intrinsic motivation systems for autonomous mental development" by Pierre-Yves Oudeyer and colleagues [128]. An early example of the use of a value system in developmental robotics can be found in the work by Xiao Huang and John Weng [235, 236].

[8] The introduction to the book *Social Motivation* by Joseph Forgas and colleagues [283] gives a good account of social motivations and their relation to cognition and affect (i.e. emotional states).

[9] To refresh your memory of affordances, refer back to the second last paragraph of Section 5.6, Chapter 5, and to Sidenote 52 in Chapter 2. Remember too that actions are focussed on the goal of the action, not on the movements by which that goal is achieved.

very effectively by crawling, and they keep using head movements when directing gaze even though eye movement alone will suffice. It isn't necessarily success at achieving task-specific goals that drives development in infants but rather the discovery of new modes of interaction with the world in which the infant is embedded: the acquisition of a new way of doing something through exploration.[10]

6.1.2   Imitation

In addition to the development of skills through exploration — reaching, grasping, and manipulating what's around you — there are two other very important ways in which cognition develops. These are imitation and social interaction, including teaching. These two different drivers of development — the exploratory and the social — mirror the developmental psychology of Jean Piaget and Lev Vygotsky, respectively.[11]

Imitation[12] — the ability to learn new behaviours by observing the actions of others — is a key mechanism in development and it is innate in humans.[13]

In contrast to trial-and-error exploration and associated learning methods such as reinforcement learning (where little successes slowly guide you towards an overall strategy for accomplishing some task), imitation provides a way of learning rapidly. Note that imitation and mimicry are not the same. Imitation is focussed on learning new skills by trying to replicate by observation what someone else is doing with the results that the agent's repertoire of actions is enlarged. Mimicry, on the other hand, is simply copying, typically by using some skill that the mimicing agent already possesses.

Although present at birth, the ability to imitate develops over the first couple of years of an infant's life. Newborn infants imitate facial expressions but it is not until 15 to 18 months that an infant can imitate a variety of actions. Imitation develops in infants in four phases:

1. Body babbling;
2. Imitation of body movements;
3. Imitation of actions on objects;

[10] The view that exploration is crucial to development is supported by research findings in developmental psychology, e.g. see Claes von Hofsten's articles "On the development of perception and action" [284] and "An action perspective on motor development" [215].

[11] A paper by Kerstin Dautenhahn and Aude Billard [285] provides a succinct introduction to the differences between Jean Piaget's exploration-centred approach to infant development and Lev Vygotsky's social approach. It emphasises the importance of social interaction and teaching in development and describes a framework for development in robots. We return to Piaget and Vygotsky in Chapter 9, Section 9.6.

[12] Read Aude Billard's article "Imitation" [286] for an overview of the role imitation plays in social cognition, i.e. the way agents interact with each other cognitively, and an explanation of three ways in which imitation is modelled: theoretically based on behavioural studies, computationally based on models of neural mechanisms, and synthetically in robotics to allow robots to learn by demonstration. Andrew Meltzoff's article [287] provides an indepth overview of the importance of imitation as a fundamental component of cognitive development.

[13] Andrew Meltzoff and Keith Moore showed that newborn infants can imitate facial expressions, for example see [288, 289].

4. Imitation based on inferring intentions of others.[14]

The term *body babbling* (or sometimes *motor babbling*) derives from the more common phrase *vocal babbling* that is used to describe the seemingly random noises a baby makes before it learns to talk. The idea with body babbling is much the same: infants do not know what muscle movements are needed to achieve the goal associated with some particular action. So, to learn this, an infant engages in random trial-and-error learning. This allows the infant to generate a map between movements, or motor commands, and the resultant action. This is quickly followed by the developement of an ability to imitate body movements and gestures.

Infants who are a few months old can direct their imitation to objects around them and later on they can defer the imitation, enacting it later on, long after the imitated action has been observed.

Imitation culminates in the ability to infer the intentions of others. In this, imitation provides a bridge between different accounts of development, specifically cognitive neuroscience and developmental psychology, and how humans develop an undertanding of the intentions of other people and the ability to empathize with them.[15] This is referred to as a *Theory of Mind*, a term which is often misunderstood. It means that one agent is able to form a view (or take a perspective) on someone else's situation; it isn't meant as a theory of how cognition works in general, i.e. a theory of how a mind works. So, to have a theory of mind means to have the ability to infer what someone else is thinking and wants to do. We return to this issue of theory of mind and inferring intentions in Chapter 9, Section 9.3.

6.1.3   Development and Learning

Development and learning are related but they are not the same. We said already in this chapter that development differs from learning in that it involves the inhibition of existing abilities and that it must be able to deal with changes to the morphology or structure of the agent. In Chapter 3, we also noted that developmental systems are model generators rather than model fitters.

[14] The identification of the four phases in the development of imitation are due to Andrew Meltzoff and Keith Moore in their widely-cited article [289]. It is explained and applied in a subsequent paper by Rajesh Rao, Aaron Shon, and Andrew Meltzoff [290].

[15] Andrew Meltzoff's and Jean Decety's landmark paper [291] describes the importance of a human's innate ability to imitate, and the neural basis for imitation in the mirror neuron system (see Chapter 5, Section 5.6), in providing a mechanism for the development of a theory of mind. In their words, "In ontogeny, infant imitation is the seed and the adult theory of mind is the fruit."

We will now explain and amplify these statements.

First and foremost, development is a process which an agent undergoes to expand its repertoire of possible actions and to extend the time horizon of its capacity for prospection (i.e. the ability to anticipate events and the need to act).

Learning and development are both concerned with building models of how the world works, and how the agent fits into that model, but learning is usually based on adapting or calibrating a model provided by another agent whereas development usually entails the agent discovering that model for itself. Thus learning is focused on determining the parameters of a model provided by others, e.g. to improve performance of the correspondence between observations. On the other hand, development is focussed on generating the model in the first place: figuring out a new way of doing something or coming up with a new explanation of why something works a certain way. The model does not have to be correct: it just has to make sense for the cognitive agent.[16] Learning is often concerned with acquiring an understanding of how the world works without any reference to the agent's own perspective on matters; development is always concerned with the relationship of the agent's capabilities in the context of how the world works. Consequently, development requires two-way interaction between the agent and its environment: it involves structural coupling, to use the terminology from Chapter 5. There has to be an element of exploration and assessment, of hypothesis and testing, in the development of the cognitive agent's understanding. Furthermore, the agent's actions have to have some causal impact on its sensory perceptions. Learning can often be accomplished just on the basis of observation.

Development often involves a decrease in performance before an improvement sets in: the curve of performance often dips before rising again. We call this a *non-monotic* process. In contrast, a monotonic process or curve doesn't dip and always rises. Learning techniques are often monotonic: they usually involve continual improvement. The problem with this is that what is learned may be a good solution, but there is no guarantee that it is the best: it may not be globally optimal. In terms of learning knowledge, a monotonic learning agent can only learn new

[16] Regarding the correctness of models, the statistician George Box famously noted "Essentially, all models are wrong, but some are useful" [292].

knowledge that does not contradict what it already knows.

On the other hand, non-monotonic learning would enable the agent to learn new knowledge that is inconsistent with what it currently knows. This allows the agent to replace or override existing knowledge if it makes more sense.[17] As we noted above, in order to facilitate exploration of new ways of doing things, an agent must sometimes suspend current skills. Consequently, development differs from learning in that (a) it must inhibit existing abilities, and (b) it must be able to cater for (and perhaps cause) changes in the morphology or structure of the agent. The inhibition does not imply a loss of learned control but an inhibition of the link between a specific sensory stimulus and a corresponding motor response.

Despite these distinctions, learning and development do go hand in hand and development won't occur if the agent does not have some capacity for learning. So, with that in mind, let us discuss learning very briefly.[18]

Overall, we can distinguish three types of learning: supervised learning, reinforcement learning, and unsupervised learning.[19]

In supervised learning, the agent that is learning is provided with examples of what it needs to learn and, in particular, it can determine an error value between the correct answer and its estimate of the correct answer. These errors are vector values: they show how much the estimate differs — the magnitude of the error — as well as the direction in which it needs to adjust its estimate in order to reduce the error next time it makes that estimate. For example, if a cognitive agent is learning the relationship between the motor control values to reach for an object and the position of that object in its field of view, it will be provided with a training set that comprises pairs of correct correspondences between both data sets. Furthermore, when it tries to reach for the object, and fails, it is able to estimate not only how far away its attempt is from the correct position, but also the required adjustment in direction. Thus, in supervised learning, the teaching (or training) signals are directional errors.

Reinforcement learning[20] is also a form of supervised learning insofar as a reward signal is provided at each step of the

[17] There is an overview of monotonic and non-monotonic learning from the perspective of cognitive systems on the University of Michigan's Cognitive and Agent Architecture website [293].

[18] For an introduction to the different techniques involved in learning, see Tom Mitchell's book *Machine Learning* [294].

[19] Kenji Doya's paper "What are the computations of the cerebellum, the basal ganglia and the cerebral cortex?" [295] contains a succinct explanation of the three basic learning paradigms of supervised, reinforcement, and unsupervised learning. There is a more gentle introduction to this material in a later paper [296].

[20] For an overview of reinforcement learning, see Mance Harmon's and Stephanie Harmon's "Reinforcement Learning: A Tutorial" [297].

learning process. In this case, the teaching signals don't contain any directional information. They are simply scalar values called rewards or reinforcement signals. In the reinforcement learning paradigm, an agent takes some action which results in some change in the state of the environment (including the agent itself). It receives a reinforcement signal, or reward. Another action results in another, possibly different, reward. The goal of the learning process is to maximize the cumulative sum of the rewards over time. In this way, successful behaviour is reinforced and unsuccessful behaviour is penalized.

Unsupervised learning[21] operates with no teaching signals, just a stream of input data. The goal of learning is to uncover the statistical regularity in this input stream and, in particular, to find some mapping between the input data and the learned output that reflects the underlying order in the input data. For example, the data stream might reflect a number of different clusters; unsupervised learning provides a way of identifying and characterizing these clusters.

As you would expect, different types of development require different learning mechanisms and the human brain is good at all three types of learning, with different regions being specialized for different types.[22] Innate behaviours are honed through continuous knowledge-free reinforcement-like learning while new skills develop through a different form of learning, driven by spontaneous unsupervised play and exploration which is not directly reinforced. On the other hand, imitative learning and learning by instruction makes use of supervised learning.[23]

In summary, cognitive skills emerge progressively through the development of an agent as it learns to make sense of its world through exploration, through manipulation, imitation, and social interaction. Terry Winograd and Fernando Flores capture the essence of developmental learning, at least from the emergent perspective, in their classic book *Understanding Computers and Cognition* [87]:

> "Learning is not a process of accumulation of representations of the environment; it is a continuous process of transformation of behaviour through continuous change in the capacity of the nervous system to synthesize it. Recall does not depend on the

[21] See Zoubin Ghahramani's paper "Unsupervised Learning" [298] for a tutorial introduction to the topic.

[22] For example, Kenji Doya argues that the cerebellum is specialized for supervised learning, basal ganglia for reinforcement learning, and the cerebral cortex for unsupervised learning [295]. These regions and the learning processes are also interdependent: James McClelland and colleagues argue that the hippocampal formation and the neo-cortex form a complementary system for learning [299]. The hippocampus facilitates rapid learning of associations between events — associative learning — which is used to reinstate and consolidate learned memories in the neo-cortex in a gradual manner.

[23] For a comprehensive overview of the different approaches to supervised learning in robots, including imitation, see "A survey of robot learning from demonstration" by Brenna Argall and colleagues [300].

indefinite retention of a structural invariant that represents an
entity (an idea, image, or symbol), but on the functional ability
of the system to create, when certain recurrent conditions are
given, a behaviour that satisfies the recurrent demands or that the
observer would class as a reenacting of a previous one."

6.2   *Phylogeny* vs. *Ontogeny*

Two terms, both of which we have already encountered in Chapters 2 and 3, are important in any discussion of development and
learning. These are *phylogeny* and *ontogeny*. Phylogeny refers
to the evolution of the an agent from generation to generation
whereas ontogeny refers to the adaptation and learning of the
system *during* its lifetime. Thus, ontogeny is just another word
for development.

Development must have a starting point. This is provided by
phylogeny which determines the agents initial sensori-motor
capabilities and its innate behaviours. These are embedded in
the agent's cognitive architecture and exist at birth in a biological
cognitive agent. Ontogeny (i.e. development) then gives rise to
the cognitive capabilities that we seek.

Not all species of animal or bird are capable of significant cognitive development. In nature, there is a trade-off between the
agent's initial capabilities (its phylogenetic configuration) and its
capacity for development (its ontogenetic potential). In general,
two types of species can be distinguished: *precocial* species and
altricial species. Precocial species are those that are born with
well-developed behaviours, skills, and abilities. These are the
direct result of their genetic make-up, i.e. their phylogenetic configuration. As a result, precocial species are quite independent
at birth. Altricial species, on the other hand, are born with un-
developed behaviours and skills and they are highly-dependent
on their parents for support. However, in contrast to precocial
species, they have a much greater capacity to develop complex
cognitive skills over their life-time (*i.e.* through ontogeny).[24]

In the context of artificial cognitive system, the challenge is
to strike the right balance between precocial and altricial in the
design of a cognitive system, specifically in the design of its cog-

[24] Aaron Sloman and Jackie
Chappell argue that, rather
than viewing the precocial
vs. altricial distinction as a
simple dichotomy in phy-
logenetic configuration and
ontogenetic potential, we
should view the precocial
and altricial as two ends of
a spectrum of possible con-
figurations: "precocial skills
can provide sophisticated
abilities at birth. Altricial capabilities have the potential
to adapt to changing needs
and opportunities. So it is
not surprising that many
species have both" [301].

nitive architecture. In effect, there are two problems: to identify the innate phylogentically-endowed skills (which need not be perfect and can be tuned through some form of learning), and to establish how these capabilities are developed. Although we have already touched on these issues in Chapter 3 on cognitive architectures, here will will take the opportunity to draw more deeply on developmental psychology and neuroscience to provide some insight into these problems, specifically with a view to understanding the necessary — core — skills, capacities, and knowledge that should be present in an agent's phylogenetic configuration, i.e. its cognitive architecture, if it is to be capable of development.

6.3   Development from the Perspective of Psychology

6.3.1   The Goal-directed and Prospective Nature of Action

Evidence from many different fields of research, including psychology and neuroscience, suggests that the movements of biological organisms are organized as actions and not reactions. While *reactions* are elicited by earlier events, *actions* are initiated by a motivated subject, they are defined by goals, and they are guided by prospective information.[25] In essence, actions are organized by goals and not by their trajectories or constituent movement, although of course these matter too.

For example, when performing manipulation tasks or observing someone else performing them, subjects fixate on the goals and sub-goals of the movements not on the body parts, e.g. the hands, or the objects.[26] However, this happens only if a (goal-directed) action is implied. When showing the same movements without the context of an agent, subjects fixate the moving object instead of the goal.

Similarly, evidence from neuroscience shows that the brain represents movements in terms of actions even at the level of neural processes. For example, as we noted in Chapter 5, Section 5.6, a specific set of neurons, mirror neurons, is activated when perceiving as well as when performing an action.[27] These neurons are specific to the goal of actions and not to the mechanics

[25] The anticipatory, goal-directed, nature of action is a keystone of *A Roadmap for Cognitive Development in Humanoid Robots*, a book co-written by the author, Claes von Hofsten, and Luciano Fadiga [12].

[26] A paper by Roland Johansson and colleagues [302] describes an experiment which shows that people fixate on key landmarks such as the point where an object is grasped, the target location of object, and the support surface, but never on their hand or the moving object. This shows that gaze supports predictive motor control in manipulation and provides evidence for the prospective goal-directed nature of action we have highlighted in this chapter. A related paper by Randal Flanagan and Roland Johansson [303] shows the same behaviour when people observe object manipulation tasks.

[27] For a good overview of the mirror-neuron system, see the review article by Giacomo Rizzolatti and Laila Craighero [214].

of executing them: they are not active if there is no explicit or implied goal associated with the movement associated with the action.

Actions, as we have emphasized several times, are guided by prospective information as opposed to instantaneous feedback data. Part of the reason for this is that events may precede the feedback signals about them because in biological systems the delays in the control pathways may be substantial. If you can not rely on feedback, the only way to overcome the problem is to anticipate what is going to happen next and to use that information to control one's behaviour.

6.3.2 Core Cognitive Abilities in Infants

The phylogenetic configuration of a cognitive system provides the core from which development builds. Young infants have two core knowledge systems that provide the basis for representing objects (including persons and places) and the concept of number (numerosities).[28] While these core systems provide the foundation of cognitive flexibility, they are themselves limited in a number of ways: they are domain specific, task specific, and encapsulated (i.e. they operate fairly independently of one another).

Infants build representations of objects but only if they exhibit certain characteristics. Specifically, the entities that are considered to be object-line are complete, connected, solid bodies that maintain their identity over time, and persist through occlusion when they are hidden by other objects. Infants can keep track of multiple objects simultaneously but the number is limited to about three objects and this ability is tolerant to changes in object properties such as colour, precise shape, and spatial location.

Infants have two core systems related to numbers: one that deals with small exact numbers of objects and one that deals with approximate numbers in sets. Infants can reliably discriminate between one and two objects, and between two and three objects, but not any higher numbers.[29] The ability to quantify small numbers of items without conscious counting is called *subitization*. This capability is not dependent on modality and in-

[28] Elizabeth Spelke's article "Core Knowledge" [304] gives a very accessible overview of core knowledge systems and the way they contribute to flexible cognitive skills through development.

[29] "Core systems of number" by Lisa Feigenson, Stanislaus Dehaene, and Elizabeth Spelke [305] provides a succinct summary of both numerosity core systems: approximate representations of numerical magnitude and precise discrimination of distinct individuals.

fants can do the same with sounds. Infants also have the ability to add these small numbers. On the other hand, the approximate number system enables infants to discriminate larger sets of entities.

An important part of core knowledge has to do with people. Specifically, it as to do with interaction between infants and their carers, and the predisposition of human infants to interact with other humans. Infants are attracted by other people and are endowed with abilities to recognize them and their expressions, and to communicate with them. As we saw in Table 6.1 at the beginning of the chapter, they develop an ability to perceive the goal-directedness of the actions of other people quite quickly. Similarly, young infants exhibit a preference for the motions produced by a moving person over other motions, so-called *biological motion*.[30]

Intentions and emotions are displayed by elaborate and specific movements, gestures, and sounds that become important to perceive and control. Some of these abilities are already present in newborn infants and reflect their preparedness for social interaction. Young infants are very attracted by people, especially to the sounds, movements, and features of the human face, and they look longer at a face that directs the eyes straight at them than at one that looks to the side.[31] They also engage in some social interaction and turn-taking that among other things is expressed in their imitation of facial gestures.[32]

The ability to spatially reorient and navigate, often taken as a typical cognitive abilities, is also subject to development. While adults solve reorientation tasks by combining non-geometric information (e.g. colour) with geometric information, young children rely only on geometry.[33] Similarly, there is evidence that navigation is based on representations that are momentary rather than enduring, egocentric rather than geocentric, and limited in the information they capture about the environement.[34] When navigating, children and adults base their turning decisions on local, view-dependent, and geometry-based representations. They navigate by forming and updating a dynamic representation of their relationship to the environment. This capacity for path integration, whereby an agent navigates from point to

[30] The preference for biological motion by human infants aged four to six months was suggested in 1982 as evidence in support of the hypothesis that this is an intrinsic capacity of the human visual system [306]. Recently, in 2008, it was shown by Francesca Simion and colleagues that in fact newborn babies are sensitive to biological motion [258].

[31] A paper by Teresa Farroni and colleagues [256] describes an experiment that demonstrates that, from birth, human infants prefer to look at faces that engage them in mutual gaze, i.e. to interact with people that make direct eye contact.

[32] Two classic papers on imitation of facial features by neonates, and imitation generally, are the result of work by Andrew Meltzoff and Keith Moore [288, 289].

[33] Linda Hermer's and Elizabeth Spelke's paper [307] discusses the difference between the way young children and adults reorient in space. In contrast to adults, children rely only on geometry to reorient even when nongeometric information is available.

[34] An article by Ranxiao Wang and Elizabeth Spelke [308] discusses the characteristic dependence by humans on momentary, egocentric, and informationally-limited cues in spatial representation, contrasting it with the more widely-held and more intuitive assumption that it is based on enduring, geocentic "cognitive maps."

point, cumulatively basing the next step on the previous ones, has been found to be one of the primary forms of navigation in insects, birds, and mammals. Like other animals, humans can return to the origin of a path and travel to familiar locations along novel paths, reorienting by recognizing landmarks rather than by forming global representations of scenes.

6.3.3   Ontogeny

If the phylogenetic configuration of a cognitive system provides the core of development, then ontogeny is the path that development takes when scaffolding these abilities to generate the anticipatory, prospectively-controlled goal-directed repertoire of possible actions.

The ontogeny of a cognitive system begins with actions that are immediate and have minimal prospection, and progresses to more complex actions that bring forth increasingly prospective cognitive capabilities. This involves the development of perception-action coordination, beginning with head-eye-hand coordination, progressing through manual and bi-manual manipulation, and extending to more prospective couplings involving inter-agent interaction, imitation, and communication, gestural and vocal (refer again to Table 6.1 at the start of the chapter). This development occurs in both the innate skills with which phylogeny equips the system and in the acquisition of new skills that are acquired as part of the ontgenetic development of the system.

6.4   Conclusions

We can now make some general observations on the nature of cognitive systems that have a capacity for development.

Perhaps most important of all is the recognition that a developmental cognitive system's actions are guided by prospection, directed by goals, and triggered by affective motives. Development, then, is the process by which the system progressively extends the time-scales of its prospective capacity and extends its repertoire of actions. As a consequence, a cognitive system needs an attentional system that fixates on the goals of actions. It also needs to have some mechanism to rehearse hypothetical scenarios through internal simulation in order to predict, explain, and imagine events. There needs to be a mechanism to use the outcome of this simulation to modulate the behaviour of the system and scaffold new knowledge through generative model building. We return to this issue in the next chapter on memory.

A developmental cognitive system is capable of adaptation and self-modification, both in the sense of parameter adjustment of phylogenetic skills through learning and through the modification of the structure and organization of the system itself so that it is capable of altering its system dynamics based on experience.[35] Again, this gives it the capacity to expand its repertoire of actions, adapt to new circumstances, and enhance its prospective capabilities.

A developmental cognitive system needs an appropriate phylogenetic configuration with innate — core — abilities. This is encapsulated in its cognitive architecture. It also needs the opportunity to develop: a period of ontogeny. During this period, development is driven by both exploratory and social motives, the first concerned with the discovery of novel regularities in the world and the potential of the system's own actions, the second with inter-agent interaction, shared activities, and mutually-constructed patterns of shared behaviour. We return to this issue in the last chapter on social cognition.

[35] Although written over 60 years ago, W. Ross Ashby's classic book *Design for a Brain* [29, 30, 31] provides a very instructive analysis of the mechanisms by which this self-organizing adaptive behaviour can be achieved through homeostasis, ultrastability, and multistability. Note that there are two editions of this book and that they differ quite significantly. It is worth reading both versions to see how Ashby's thinking changed in the eight years that elapsed between the publication of the first and second editions. See an open peer commentary article by the present author for a brief and selective overview of the differences between the two editions.

7
Memory and Prospection

7.1 Introduction

We began our study of artificial cognitive systems in Chapter 1 with a general overview of the nature of cognition, highlighting the essential characteristics of a cognitive system. We focussed in particular on the ability of a cognitive agent to anticipate the need for action and expand its repertoire of actions. We saw that cognitive systems develop and learn, and in so doing adapt to changing circumstances. In Chapter 2, we looked at the different paradigms of cognitive science and the different assumptions people make about cognition. As we saw, this has a significant impact on the way that cognitive systems are modelled. Then, in Chapter 3, we looked at various aspects of these models under the general heading of cognitive architectures. One of the central issues in all of the cognitive architectures we studied was concerned with what a cognitive systems knows and how it acquires, retains, and uses its knowledge and know-how, i.e. its cognitive skills. This brings us to our next topic: memory.

Memory plays a crucial and sometimes unexpected role in cognition. In this chapter, we ask what a cognitive system remembers, why it remembers, and how it uses what it remembers. In answering these questions we will build on what we learned in Chapters 4, 5, and 6 on autonomy, embodiment, and development & learning, respectively.

There is a strong parallel between memory and knowledge and, in the following, we will examine this relationship. It is tempting to think of memory just as a passive mechanism for storing knowledge and to focus exclusively on knowledge. As we saw in Chapters 2 and 3, knowledge is central to cognitivist cognitive systems, providing the content that complements the cognitive architecture. Together, they form the complete cognitive model. Very often, as a natural consequence of cognitivism's close relationship with classical AI and the physical symbol system hypothesis in particular (see Section 2.1.2), knowledge is assumed to be symbolic. However, this assumption may not always hold. So, to avoid unintentional misunderstandings about the nature of knowledge, in this chapter we will tackle the issue from the perspective of memory, highlighting the similarities and, often, the duality of memory and knowledge. We will not view knowledge as simply the contents of memory and we will view memory and knowledge as being in some sense equivalent in that they both encapsulate the experience that arises from interaction with the world.

This is how we will proceed. We will begin with the simplest task: distinguishing between different types of memory. We will differentiate between several types of memory, including declarative, procedural, semantic, episodic, long-term, short-term, working, modal, amodal, symbolic, sub-symbolic, hetero-associative, and auto-associative memory. We have already met many of these distinctions in Chapter 3 when discussing cognitive architectures. Our goal here is to explain the differences.

Once we are aware of the different types of memory, we will discuss the role of memory. It will quickly become clear that memory has as much to do with the future as it does with the past. Memory facilitates the persistence of knowledge and forms a reservoir of experience. Without it, it would be impossible for the system to learn, develop, adapt, recognize, plan, deliberate, and reason. Memory functions to preserve what has been achieved through learning and development, ensuring that, when a cognitive systems adapts to new circumstances, it doesn't lose its ability to act effectively in situations to which it had adapted previously. But memory has another role in addition to preserving past experience: it is to anticipate the future. In this context, we will discuss one of the central pillars of cognitive capacity: the ability to simulate internally the outcomes of possible actions and select the one that seems most appropriate for the current situation. Viewed in this light, memory can be seen as a mechanism that allows a cognitive agent *to prepare to act*, overcoming through anticipation the inherent "here-and-now" limitations of its perceptual capabilities.

7.2   *Types of Memory*

7.2.1   *Overview*

We will begin by differentiating between different types of memory.[1] As we noted in the previous section, there are many distinctions to be made. These include the following.

- short-term *vs.* long-term
- declarative *vs.* procedural
- semantic *vs.* episodic
- symbolic *vs.* sub-symbolic
- modal *vs.* amodal modal
- hetero-associative *vs.* auto-associative

That isn't the end of the story, though. Declarative memory is sometimes referred to as propositional or descriptive memory. Short-term memory is sometimes referred to as working memory but there are some subtle distinctions to be made here too. To give us some basis for making these distinctions, it may be helpful to first consider what memory is. To a large extent, memory can be viewed as something that results from past experience and which is available for recall to support effective on-going and future behaviour. We are being a little evasive here by referring to it as "something." There is a reason for this evasiveness and it goes back (once again) to the differences between cognitivist and emergent approaches to cognition.

If you adhere to the cognitivist approach, you would call that something knowledge. If you adhere to the emergent approach, you would be less willing to commit to this term because it might be misconstrued to mean that this "something" is a neatly encoded and probably symbolic description — or representation — of the world that the cognitive agent has experienced.

[1] For an overview of memory in both natural and artificial cognitive systems, see the review by Rachel Wood, Paul Baxter, and Tony Belpaeme [309]. Larry Squire's article "Memory systems of the brain: A brief history and current perspective" [310] provides a succinct and accessible overview from a neuroscientific perspective.

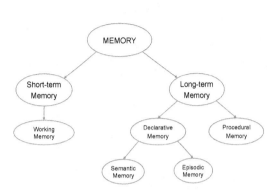

Figure 7.1: A simplified taxonomy of memory. For a more complete taxonomy, see [310].

Someone who adheres to the emergent paradigm would prefer to think of memory as some state of the cognitive system that can be recalled in the service of current, imminent, or future action. We called this action *effective* above and we should remind ourselves again what we learned in Chapters 1 and 2: that *effective* means adaptive and anticipatory, so that a cognitive system doesn't operate just on the basis of its current sensory data but readies itself for what it expects and adjusts to the unexpected.

One further point: in contrast with most artificial computer-based systems, memory in natural systems is not localized and it is not passive. On the contrary, it is distributed and active to the extent that memory should be thought of not as a "storage location" in a cognitive architecture, but a pervasive facet of the complete cognitive system, fully integrated into all aspects of the cognitive architecture. In this sense, memory is an active process — a primary mechanism of cognition — rather than a passive information repository. Increasingly, this position is being adopted in artificial cognitive systems.[2]

7.2.2 Short-term and Long-term Memory

Let us now look at different types of memory (refer to Figure 7.1 as you read through this section).[3] The first distinction we

[2] On the pervasive view of memory in a cognitive system, Joaquín Fuster remarks "We are shifting our focus from 'systems of memory' to the memory of systems. The same cortical systems that serve us to perceive and move in the world serve us to remember it." [311]. An example of how this perspective is being applied in the design of artificial cognitive systems can be found in Paul Baxter's and Will Browne's Memory-Based Cognitive Framework (MBCF) [312].

[3] The overview of the different types of memory follows closely the treatment in "A review of long-term memory in natural and synthetic system" by Rachael Wood, Paul Baxter, and Tony Belpaeme [309].

make is between *short-term memory* and *long-term memory*. The difference between them is evidently based on how long the content of memory lasts but there are other differences to do with the physical process by which the memory is retained. In short-term memory, the memory is maintained by transitory electrical activity while long-term memory uses longer-lasting chemical changes in the neural system.[4] Short-term memory is sometimes referred to as working memory, a form of temporary memory that is used to support current cognitive processing, such as the achievement of goal-directed action.[5] We already met this in Chapter 3 where we discussed the Soar and ISAC cognitive architectures (see Sections 3.4.1 and 3.4.3).

7.2.3 Declarative and Procedural Memory

We can also distinguish memory based on the nature of what is remembered and the type of access we have to it. Specifically, memory can be categorized as either *declarative* or *procedural*, depending on whether it captures knowledge of things — facts — or actions — skills. Sometimes they are characterized as memory of knowledge and know-how: "knowing that" and "knowing how."[6] This distinction applies mainly to long-term memory but short-term memory too has a declarative aspect. Declarative memory is sometimes referred to as *propositional memory* because it refers to information about the agent's world that can be expressed in the form of propositions. This is significant because propositions are either true or false. Thus declarative memory typically deals with factual information. This is not the case with skill-oriented procedural memory. As a consquence, declarative memories, in the form of knowledge, can be communicated from one agent to another through language, for example, whereas procedural memories can only be demonstrated.

Declarative and procedural memory differ in other ways. Declarative knowledge can be acquired in a single act of perception or cognition whereas procedural memories are acquired progressively and may require an element of practice. Declarative memory is accessible to conscious recall but procedural memory is not. You can engage in introspection about your

[4] Rachael Wood and colleagues [309] note that the distinction between short- and long-term memory was first made by Donald Hebb in 1949 in his book *The Organization of Behaviour* [61]. On the other hand, the review by Nelson Cowan [313] puts forward the view that short-term memory is an activated part of long-term memory. Joaquín Fuster supports this position, viewing working memory as the temporary activation of short-term and long-term perceptual and motor memory [311].

[5] For an overview of neurocomputational models of working memory, see [314]. The authors, Daniel Durstewitz, Jeremy Seamans, and Terrence Sejnowski, refer to working memory as a form of short-term memory, again making the point that short-term memory depends on the maintenance of elevated activity (i.e. firing rates) in sub-populations of neurons in the absence of external cues. This contrasts with more long-lasting chemical changes in the neural synapses typical of long-term memory. The article reviews the different ways that this persistent elevated activity is achieved.

[6] The distinction between *knowing that* and *knowing how* was made in 1949 by Gilbert Ryle in his book *The Concept of Mind* [315].

declarative memories but you can't think in this way about your skills and motor abilities. Thus, an agent has conscious access to its declarative memory of facts but skills and motor abilities are accessed unconsciously. For this reason, declarative and procedural memory are sometimes referred to as *explicit* and *implicit* memory, respectively.

The term *non-declarative memory* is sometimes applied to procedural memory. However, in general, this term also embraces other forms on non-conscious memory.[7]

7.2.4   *Episodic and Semantic Memory*

Two different types of declarative memory can be distinguished. These are *episodic memory* and *semantic memory*.[8]

Episodic memory refers to specific instances in the agent's experience while semantic memory refers to general knowledge about the agent's world which may be independent of the agent's specific experiences. In this sense, episodic memory is autobiographical. By its very nature in encapsulating some specific event in the agent's experience, episodic memory has an explicit spatial and temporal context: what happened, where it happened, and when it happened. This temporal sequencing is the only element of structure in episodic memory. Episodic memory is a fundamentally *constructive* process.[9] Each time an event is assimilated into episodic memory, past episodes are reconstructed. However, they are reconstructed a little differently each time. This constructive characteristic is related to the role that episodic memory plays in the process of internal simulation that forms the basis of prospection, the key anticipatory function of cognition (see Section 7.4 below).

In contrast, semantic memory "is the memory necessary for the use of language. It is a mental thesaurus, organized knowledge a person possesses about words and other verbal symbols, their meaning and referents, about relations among them, and about rules, formulas, and algorithms for the manipulation of the symbols, concepts, and relations."[10] While this definition of semantic memory dates from 1972, it is still valid today, especially in the cognitivist paradigm of cognitive science. It also

[7] Non-declarative procedural memory is subdivided into four types: skills and habits, priming, classical conditioning, and non-associative learning. For more details, see Larry Squire's article "Memory systems of the brain" [310].

[8] The term *episodic memory* was coined by Endel Tulving in 1972 in an article entitled 'Episodic and Semantic Memory'" [316]. His 1983 book *Elements of Episodic Memory* is his definitive work and his overview of the book, *Précis of Elements of Episodic Memory* [317] is essential reading, not only for its clear explanation of the many differences between episodic memory and semantic memory, but for its characterization of declarative (propositional) memory and procedural memory. As we will see, episodic memory plays a key role in cognition and in the anticipatory aspect of cognition in particular.

[9] The constructive characteristic of episodic memory is emphasized by Martin Seligman and colleagues in their paper on prospection "Navigating into the Future or Driven by the Past" [318].

[10] This quotation explaining the characteristics of semantic memory appears in Endel Tulving's 1972 article [316], p. 386 and is quoted in his *Précis* [317].

explains the linguistic origins of the term.

Episodic memory and semantic memory differ in many ways. In general, semantic memory is associated with how we understand (or model) the world around us, using facts, ideas, and concepts. On the other hand, episodic memory is closely associated with experience: perceptions and sensory stimulus. While episodic memory has no structure other than its temporal sequencing, semantic memory is highly structured to reflect the relationships between constituent concepts, ideas, and facts. Also, the validity (or truth: remember semantic memory is a subset of propositional declarative memory) of semantic memories is based on social agreement rather than personal belief, as it is with episodic memory.[11]

Semantic memory can be derived from episodic memory through a process of generalization and consolidation. Episodic memory can be both short-term and long-term while semantic and procedural memory are long-term.

In artificial cognitive systems, declarative memory is usually based on symbolic information. On the other hand, episodic memory and procedural memory often exploit non-symbolic information, sometimes referred to as sub-symbolic memory.

7.2.5 Modal and Amodal Memory

In Section 5.10 we encountered the distinction between modal and amodal representation in our discussion of grounded cognition. Modal memory is tied directly to a particular sensory modality such as vision, audition, or touch. On the other hand, amodal memory has no necessary association with the sensorimotor experiences. Semantic declarative facts, represented symbolically, are typically amodal, especially considered from the perspective of the cognitivist paradigm. Episodic memory though is more likely to be modal since it is closely tied to an agents's specific experiences.

7.2.6 Auto- and Hetero-Associative Memory

In Section 2.2.1 we discussed associative memory, noting that there are two variants: hetero-associative and auto-associative.

[11] Semantic memory and episodic memory can be contrasted in many other ways: twenty-seven differences are listed in [319], p. 35. Interestingly, this article notes the clear applicability of semantic memory in artificial intelligence but questions the relevance of episodic memory for AI. Today, the importance of episodic memory is widely accepted, even in the cognitivist community, as one can see by its relatively recent incorporation in the Soar cognitive architecture (see [113]).

The main idea with associative memory is that some element of information or, more generally, some pattern is associated with and linked to another element of information or pattern. The first element or pattern is used to recall the second, by association. Hetero-associative memory recalls a memory that is different in character from the input; a particular smell or sound, for example, might evoke a visual memory of some past event. On the other hand, auto-associative memory recalls a memory of the same modality as the one that evoked it: picture of a favourite object might evoke a mental image of that object in vivid detail.

7.3   The Role of Memory

Why do we remember things? One answer is so that we can recognize objects, events, and people we've encountered before and act towards them in some appropriate way: avoiding the things that we have discovered are dangerous but not the things that are good for us. Putting it another way: memory is what makes it possible for the changes that occur as a result of learning and development to persist. That is clearly very important. However, there is another and possibly more important role of memory in cognitive systems. It is not to remember, but to anticipate.[12] The White Queen in Lewis Carroll's classic *Through the Looking Glass* puts it nicely when she tells Alice that "It's a poor sort of memory that only works backwards."[13] It is an important observation and one that resonates with our comments above that memory is increasingly coming to be understood not simply as a repository of past experiences and learned knowledge, but as an active and pervasive cognitive mechanism in its own right. One might even view memory as the engine that drives cognition, especially when you consider it as a way of looking forwards rather than backwards.[14]

There are a number of implications that arise from this forward-looking perspective on memory.[15] The first is that memory is an active process, as we have said, and also that it is fundamentally associative. Memories are recalled by associated triggers, which of course, can be other memories. If you have a network of associative memories, you can run through this network backwards

[12] Alain Berthoz sums up the purpose of memory like this: "Memory is used primarily to predict the consequences of future action by recalling those of past action" [320]. Daniel Schacter's and Donna Addis's essay in Nature, "Constructive memory — The ghosts of past and future" explains the way memories flexibly recombine to anticipate the future [321].

[13] See Chapter 5 of Lewis Carroll's *Through the Looking Glass* [322].

[14] The brain is geared up to anticipate. Keith Downing's "Predictive Models in the Brain" [323] highlights five different predictive architectures in the different regions of the brain: the cerebellum, basal ganglia, hippocampus, neo-cortex, and the thalamocortical system. In line with the distinction we drew above between declarative and procedural memory, the first two deal with procedural prediction whereas the last three are more concerned with declarative prediction.

[15] The associative and developmental implications of active memory-based cognition are discussed in Paul Baxter's and Will Browne's paper "Memory as the substrate of cognition: a developmental robotics perspective" [312].

or forwards. Running through it forwards provides the anticipatory predictive element of memory suggesting possible sequence of events leading to a desired goal. Running through it backwards provides a way of explaining how some event or other might have occurred[16] or imagining ways in which it might have turned out differently.[17] The second implication is that cognition is inherently a developmental process. Memory, and memories, are both the process and the product of interaction with the world and memory reflects the way that a cognitive agent comes to understand that world in a way that facilitates these interactions. This associative prospective-retrospective view of memory is more in tune with the emergent paradigm of cognitive systems than it is with the cognitivist paradigm.

7.4   Self-projection, Prospection, and Internal Simulation

Memory plays at least four roles in cognition: it allows us to remember past events, anticipate future ones, imagine the viewpoint of other people, and navigate around our world. All four involve *self-projection*: the ability of an agent to shift perspective from itself in the here-and-now and to take an alternative perspective. It does this by *internal simulation*, i.e. the mental construction of an imagined alternative perspective.[18] Thus, there are four forms of internal simulation:[19]

1. Episodic memory (remembering the past).
2. Navigation (orienting yourself topographically, i.e. in relation to your surroundings).
3. Theory of mind (taking someone else's perspective on matters).
4. Prospection (anticipating possible future events).

Each form of simulation has a different orientation (past, present, or future) and each refers to the perspective of either the first person, i.e. the agent itself, or another person.

We have already met the concept of *theory of mind* in Chapter 6, Section 6.1.2, and we will return to it again in Chapter 9, Section 9.3. We discussed navigation briefly in Chapter 6, Section 6.3.2, when describing the core abilities that provide the founda-

[16] The explanatory process is sometimes called *abduction* or *abductive inference*.

[17] Imagining different outcomes is referred to as *counterfactual thinking*; literally, counter to the facts.

[18] For an overview of the concept of *simulation* in cognitive psychology and neuroscience, read "Episodic Simulation of Future Events — Concepts, Data, and Applications" by Daniel Schacter, Donna Addis, and Randy Buckner [324].

[19] The four types of self-projection through mental simulation — episodic memory, navigation, theory of mind, and prospection — are proposed by Randy Buckner and Daniel Carroll in their article "Self-projection and the Brain" [325].

tion for learning and development. In the next section, we will focus on prospection in relation to episodic memory.[20]

Recent evidence suggests that all four kinds of internal simulation involve a single core brain network and this network overlaps what is known as the *default-mode network*, a set of interconnected regions in the brain that is active when the agent is not occupied with some attentional task.[21]

It is significant that all four forms of simulation are constructive, i.e., they involve a form of imagination. This may not seem odd in the case of prospection, theory of mind, or even navigation, but it does seem curious in the context of remembering the past. However, when we engage in retrospection, we don't just try to recall events, but very often we reconstruct events to see how they could have turned out differently. Later in this chapter, we will explain further why it is believed that memory, and episodic memory in particular, is constructive and not simply a store of perfect recollections.

We now take a closer look at the link between episodic memory and prospection, as well as the role of affect and emotion in prospection. Section 7.5.8 then takes up the issue of internal simulation in the context of action and motor control.

7.4.1 Prospection and Episodic Memory

As we have emphasized throughout this book, anticipation is one of the central characteristics of cognition. While retrospection refers to the ability to re-experience the past, prospection refers to the brain's ability to experience the future by simulating what it might be like. To a large extent, cognition is the mechanism by which we prepare to act and without which we — or an artificial cognitive system — would be ill-equipped to deal with the uncertain constantly-changing and often precarious conditions of our environment.

There is a difference between knowing about the future and projecting ourselves into the future. The latter is experiential and the former is not. Thus, as you might suspect, episodic memory (memory of experiences) and semantic memory (memory of facts) facilitate different types of prospection. Episodic mem-

[20] For an overview of the nature of prospection — the mental simulation of future possibilities — and the central role it plays in organizing perception, cognition, affect, memory, motivation, and action, you should read "Navigating into the Future or Driven by the Past" by Martin Seligman, Peter Railton, Roy Baumeister, and Chandra Sripada [318]. Randy Buckner and Daniel Carroll note that prospection and related concepts are referred to in various ways, e.g. *episodic future thinking, memory of the future, pre-experiencing, proscopic chronesthesia, mental time travel,* and just plain *imagination.* They also remind us that prospection can involve conceptual content and affective — emotional — states. We say more about this in Section 7.4.2.

[21] Evidence for the involvement of the *default-mode network* in remembering the past and envisioning the future, i.e. in prospection, is reported by Ylva Østby and colleagues in "Mental time travel and default-mode network functional connectivity in the developing brain" [326].

ory allows you to re-experience your past and pre-experience your future. There is evidence that projecting yourself forward in time is important when you form a goal, creating a mental image of yourself acting out the event and then episodically pre-experiencing the unfolding of a plan to achieve that goal.[22]

We mentioned already that episodic memory is inherently constructive: old episodic memories are reconstructed slightly differently every time a new episodic memory is assimilated or remembered. Why would this be so? While episodic memory certainly needs some constructive capacity to assemble individual details into a coherent memory of a given episode, the *constructive episodic simulation hypothesis*[23] suggests that its role in prospection involving the simulation of multiple possible futures imposes an even greater need for a constructive capacity because of the need to extrapolate beyond past experiences. In other words, simulating multiple yet-to-be-experienced futures requires flexibility in episodic memory. This flexibility is possible because episodic memory is not an exact and perfect record of experience but one that conveys the essence of an event and is open to re-combination.

It is worth noting that the ability to pre-experience the future does not appear in human children until the third or fourth year of life, very late in the child's overall development, and much later than other cognitive abilities.[24]

7.4.2   Prospection and Affect

When humans imagine the future, they not only anticipate an event, they also anticipate how they feel about that event. They do so for a very good reason: knowing how you feel about something is a very good way of telling whether or not that event is safe or dangerous. We call these the *hedonic* consequences of the event: whether we feel good about it or bad about it, whether it is associated with pleasure or pain, lack of concern or fear. Thus, the pre-experience of prospection also involves "pre-feeling." The brain accomplishes prospection by simulating the event and the associated hedonic experience.[25] However, pre-feeling is not always reliable because contextual factors also play a part in the

[22] This use of episodic memory in prospection is referred to as *episodic future thinking*, a term coined by Cristina Atance and Daniela O'Neill to refer to the ability to project oneself forward in time to pre-experience an event. Their article [327] explains the role of episodic memory in prospection and its relationship to semantic memory. For an overview, see Karl Szpunar's review "Episodic Future Thought: An Emerging Concept" [328].

[23] The constructive episodic simulation hypothesis was proposed by Daniel Schacter, Donna Addis, and colleagues: refer their seminal articles [324, 329] for more details and also see a review in [328].

[24] Cristina Atance and Daniela O'Neill established that episodic future thinking ability appears between 3 and 4 years of age; see their article "The emergence of episodic future thinking in humans" [330],

[25] For an overview of the importance of hedonic experience in prospection, and its short-comings, read "Prospection: Experiencing the Future" by Daniel Gilbert and Timothy Dawson [331]. It explains the idea of "pre-feeling," i.e. the hedonic experience during simulation and it explains why prefeelings are not always reliable predictors of subsequent hedonic experiences: you don't always end up feeling the way you thought you would feel about a future event.

hedonic experience.

In general, a pre-feeling will be a good predictor of subsequent hedonic experiences only if (a) the influence of a simulation of an event on our current hedonic experience is the same as the eventual perception of that event on our future hedonic experience, and (b) the current contextual factors are the same as the future contextual factors. Errors in prospection arise when either of these two conditions are not met. Many of the errors are due to inadequacies in the simulation.

There are four types of problems with simulation in humans. First, simulations can be unrepresentative. We don't always use the most appropriate memories to imagine the future event, often using an extreme memory of a past event (either bad or good) to imagine such an event in the future. Second, simulations are based on memories that retain only the essentials of the event. The problem is that the non-essential elements often have a significant impact on subsequent hedonic experience. Consequently, people tend to predict that good events will be better in the future and, *vice versa*, bad events will be worse. Third, simulations are abbreviated and are focussed on the early aspects of an event: they over-emphasize the initial moments of the event. As a result, these simulations under-estimate how quickly we adapt and therefore they don't represent how we will actually feel about an event.

These three problems with simulation interfere with meeting condition (a) above so that the influence of the simulation on the pre-feeling is not the same as the influence of the eventual perception on the eventual feeling. The fourth problem with simulations is that they are decontextualized: they don't reflect the contextual conditions that can have a significant impact on hedonic experience. This interferes with condition (b) and causes people to predict future hedonic experience inaccurately.

Why we are spending so much time discussing feeling in a book devoted to cognition? The reason is that feeling — affect or emotion — plays a pivotal role in cognitive behaviour, influencing the decisions we make and the actions we select.[26] Cognition is not just about rational analysis. It is as much about acting effectively, as we have stated previously. The preceding discussion

[26] For a detailed review of the link between affect and action from several perspectives — psychological, neurophysiological, and computational — see [332]. The authors, Rob Lowe and Tom Ziemke, also put forward the view that emotional feelings are predictors of action tendency, i.e. the actions that an agent is primed to take. In their model, these predictions either increase or decrease the action tendency in relation to feedback from the body (as well as from the perceived, e.g. social, context). Thus, emotional feelings and actions combine in a single system of predictive regulation and feedback in a manner that is similar to allostasis (see Section 4.3.4).

For a more general overview of the relevance of emotion to cognition, read Rosalind Picard's article "Affective Computing" [333] or her book of the same title [334]. The term *affective computing*, coined by Picard in 1995, refers both to the role of emotion in modulating behaviour and decision making and to the inference of a person's emotional state by interpreting visual and aural cues. More broadly, affective computing includes the design of computers that can detect, respond to, and simulate human emotional states (e.g. see [335]). This is an important part of an artifical cognitive system's ability to infer the intentions of a human and predict how they will act, a topic we discuss in more detail in Chapter 9.

of the relationship between hedonic experience and prospection shows the impact that feeling can have on the way people assess a future situation (sometimes incorrectly) and consequently on the way they will act. We have come across this relationship between affect and cognition several times already in the book. We remarked on the affective aspect of homeostasis in Chapter 4, Section 4.3.3, and the link between affect and embodiment in Chapter 5, Sections 5.6 and 5.7. Affect is also an important consideration when you address the internal value system that provides the drives and motivations underpinning learning and development, as we saw in Chapter 5, Section 5.9, and Chapter 6, Section 6.1.1.

7.5   Internal Simulation and Action

So far, we have considered internal simulation entirely in terms of memory-based self-projection, using re-assembled combinations of episodic memory to pre-experience possible futures, re-experience (and possibly adjust past experiences), and project ourselves into the experiences of others. However, we know from Chapter 5 that action plays a significant role in our perceptions so the question then is: does action play a role in internal simulation? The answer is a clear 'yes'.[27] Internal simulation extends beyond episodic memory and includes simulated interaction, particularly embodied interaction. Although the terms simulation, internal simulation, and mental simulation are widely used, you will also see references being made to *emulation*, very often when the approach endeavours to model the exact mechanism by which the simulation is produced.[28]

7.5.1   The Simulation Hypothesis

There are a number of simulation theories, but perhaps the most influential is what is known as the *simulation hypothesis*.[29] It makes three core assumptions:

1. The regions in the brain that are responsible for motor control can be activated *without* causing bodily movement.

[27] The literature on embodied internal simulation, i.e. internal simulation that binds perception and action together, is extensive. To get started, read Germund Hesslow's articles on the *simulation hypothesis* [100, 336]. Then read Henrik Svensson's, Jessica Lindblom's, and Tom Ziemke's paper [218] for a more detailed overview of various approaches to simulation.

[28] Rick Grush, for example, when describing processes similar to those outlined in this section, uses *emulation* to distinguish his theory from alternative simulation theories that do not incorporate a model of the effect that off-line motor commands would have on the agent's perceptions [99].

[29] The *simulation hypothesis* was first put forward by Germund Hesslow in 2002 [100]. His recent review [336] revisits the hypothesis, provides additional neuroscientific evidence, and considers its implications.

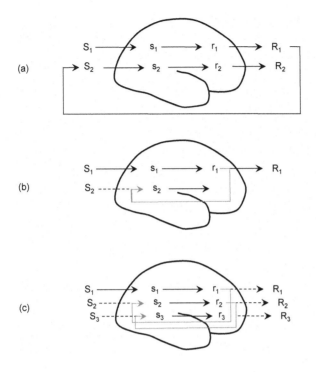

Figure 7.2: Internal simulation. (a) stimulus S_1 elicits activity s_1 in the sensory cortex. This leads to the preparation of a motor command r_1 and an overt response R_1. This alters the external situation, leading to S_2, which causes new perceptual activity, and so on. There is no internal simulation. (b) The motor command r_1 causes the internal simulation of an associated perception of, for example, the consequence of executing that motor command. (c) The internally simulated perception elicits the preparation of a new motor command r_2, i.e. a covert action, which in turn elicits the internal simulation of a new perception s_3 and a consequent covert action r_3, and so on. From [100], © 2002, with permission from Elsevier.

2. Perceptions can be caused by internal brain activity and not just by external stimuli.

3. The brain has associative mechanisms that allow motor behaviour or perceptual activity to evoke other perceptual activity.

The first assumption allows for simulation of actions and is often referred to as *covert action* or *covert behaviour*. The second allows for simulation of perceptions. The third assumption allows simulated actions to elicit perceptions that are like those that would have arisen had the actions actually been performed. There is an increasing amount of neurophysiological evidence in support of all three assumptions.[30]

If we link these assumptions together, we see that the simulation hypothesis shows how the brain can simulate extended perception-action-perception sequences by having the simulated perceptions elicit simulated action which in turn elicits simulated

[30] For a summary of the neuroscientific evidence in support of the simulation hypothesis, refer to a paper by Henrik Svensson and colleagues on dream-like activity in the development of internal simulation [337].

perceptions, and so on. Figure 7.2 summarizes the simulation hypothesis, showing three situations, one where there is no internal simulation, one where a motor response to an input stimulus causes the internal simulation of an associated perception, and one where this internally simulated perception then elicits a covert action which in turn elicits a simulated perception and a consequent covert action, and so on.

7.5.2   Motor, Visual, and Mental Imagery

Action-directed internal simulation involves three different types of anticipation: implict, internal, and external.[31] *Implicit anticipation* concerns the prediction of motor commands from perceptions (which may have been simulated in a previous phase of internal simulation). *Internal anticipation* concerns the prediction of the proprioceptive consequences of carrying out an action, i.e. the effect of an action on the agent's own body. *External anticipation* concerns the prediction of the consequences for external objects and other agents of carrying out an action.[32] Implicit anticipation selects some motor activity (possibly covert, i.e. simulated) to be carried out based on an association between stimulus and actions; internal and external anticipation then predict the consequences of that action. Collectively, they simulate actions and the effects of actions.

Covert action involves what is referred to as *motor imagery* and simulation of perception is often referred to as *visual imagery*. Perceptual imagery would perhaps be a better term since there is evidence that humans use imagery from all the senses. In a way, motor imagery is also a form of perceptual imagery, in the sense that it involves the proprioceptive and kinesthetic sensations associated with bodily movement. However, reflecting the interdependence of perception and action, covert action often has elements of both motor and visual imagery and, *vice versa*, the simulation of perception often has elements of motor imagery. Visual and motor imagery are sometimes referred to collectively as *mental imagery*.[33] As such, mental imagery can be viewed as a synonym for internal simulation.[34]

[31] The three functional parts of the simulation process (implicit, internal, and external anticipation) are described in a paper by Henrik Svensson and colleagues: "Representation as Internal Simulation: A Minimalistic Robotic Model" [338].

[32] The terms *internal anticipation* and *external anticipation* are also referred to as *bodily anticipation* and *environmental anticipation* [337].

[33] For an overview of *mental imagery* from the perspective of psychology, see Samuel Moulton's and Stephen Kosslyn's article [339]. They identify several different types of perceptual imagery and distinguish between two different types of simulation: *instrumental simulation* and *emulative simulation*. The former concerns itself only with the content of the simulation while the latter also replicates the process by which that content is created in the simulated event itself. They refer to this as *second-order simulation*. For a computational perspective on mental imagery, specifically in the context of cognitive architecture, see Samuel Wintermute's survey [340].

[34] Samuel Moulton and Stephen Kosslyn put it like this: "all imagery is mental emulation" [339], p. 1276.

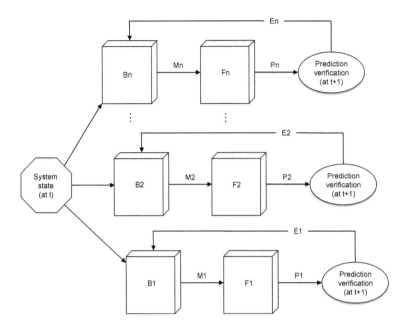

Figure 7.3: The HAMMER architecture, showing multiple inverse models (B1 to Bn) taking as input the current system state, which includes a desired goal, suggesting motor commands (M1 to Mn), with which the corresponding forward models (F1 to Fn) form predictions of the system's next state (P1 to Pn). These predictions are verified at the next time state, resulting in a set of error signals (E1 to En). From [232], © 2006, with permission from Elsevier. See also [233] for an alternative rendering of the HAMMER architecture.

7.5.3   Internal Simulation in Artificial Cognitive Systems

While internal simulation is an essential aspect of human cognition, it is also an increasingly-important part of artificial cognitive systems, as we saw in the ISAC cognitive architecture in Chapter 3, Section 3.4.3. We also discussed internal simulation in Chapter 5, Section 5.8, on Off-line Embodied Cognition, in general, and in the HAMMER architecture, in particular.[35] We will now look a little closer at how the HAMMER architecture effectively builds on the simulation hypothesis.

Recall that HAMMER accomplishes internal simulation using forward and inverse models which encode internal sensorimotor models that the agent would utilise if it were to execute that action (see Figure 7.3).

The inverse model takes as input information about the current state of the system and the desired goal, and it outputs the motor commands necessary to achieve that goal.

The forward model acts as a predictor. It takes as input the motor commands and simulates the perception that would arise

[35] HAMMER (Hierarchical Attentive Multiple Models for Execution and Recognition) is an architecture for internal simulation developed by Yiannis Demiris and Bassam Khadhouri [232, 233]. We discussed it briefly in Chapter 5, Section 5.8 and Sidenote 50.

if this motor command were to be executed, just as the simulation hypothesis envisages. However, the HAMMER architecture takes internal simulation one step further by providing the output of the inverse model as the input to the forward model. This allows a goal state (demonstrated, for example, by another agent or possibly recalled from episodic memory) to elicit the simulated action required to achieve it. This simulated action is then used with the forward model to generate a simulated outcome, i.e. the outcome that would arise if the motor commands were to be executed. The simulated perceived outcome is then compared to the desired goal perception and the results are then fed back to the inverse model to allow it to adjust any parameters of the action.

A distinguishing feature of the HAMMER architecture is that it operates multiple pairs of inverse and forward models in parallel, each one representing a simulation — a hypothesis — of how the goal action can be achieved. The choice of inverse/forward model pair is made by an internal attention process based on how close the predicted outcome is to the desired one. Furthermore, it provides for the hierarchical composition of primitive actions into more complex sequences and it has been implemented both in robot simulations and on physical robotic platforms.

7.5.4   *Episodic and Procedural Memory in Internal Simulation*

The forward and backward models of the HAMMER architecture reflect the bi-directional nature of memory that we noted at the beginning of this chapter. What is significant here is that we have now gone beyond the scope of episodic memory in effecting internal simulation by invoking actions and behaviours. The sensorimotor associations involved in internal simulations, for forward and inverse models, requires both episodic memory and procedural memory. Episodic memory is needed for visual imagery, including proprioceptive imagery, whereas procedural memory is needed for motor imagery.

Classical treatments of memory, such as the way we described it in Section 7.2 above, usually maintain a clear distinction between declarative memory and procedural memory, in general, and between episodic memory and procedural memory, in particular. However, contemporary research takes a slightly different perspective, binding the two more closely. We already saw an example of this in Section 5.6 on the interdependency of action and perception and on the mirror neuron system, in particular. While it is still a major challenge to understand how these two memory systems are combined, this coupling is the basic idea underpinning joint perceptuo-motor representations[36] We will develop this issue further in the next chapter in Section 8.5. For now we finish our discussion of internal simulation by addressing imagination.

[36] Examples of joint perceptuo-motor techniques include Marco Iacoboni's instantiation of *Ideo-motor Theory* [341] and *Theory of Event Coding* by Bernhard Hommel and colleagues [342].

7.5.5 *Functional Imagination*

At the beginning of Section 7.4, we referred informally to imagination. We will now bring a little more structure to our understanding of this term.

Generally speaking, imagination refers to cognitive activity that operates without direct recourse to an agent's sensory system. That's exactly what we have been discussing here with respect to internal simulation and imagination loosely associated with visual and motor imagery. We can be a little more specific though if we refer to *functional imagination:*[37] a mechanism that allows an agent to simulate its own actions and behaviours, predicting their sensorimotor consequences for some operational advantage. The advantage could be a reward such as finding a source of power or it could be some systemic change linked to the agent's value system, drives, or affective motivations, such as those we discussed in Chapter 6.

[37] The concept of *functional imagination* was suggested by Hugo Gravato Marques and Owen Holland [343] along with five necessary and sufficient conditions for its implementation.

Five conditions have been proposed for implementing an artificial cognitive system capable of functional imagination. They are:

1. An agent must be able to predict the sensory consequences of motor actions.

2. An agent must be able to represent the sensory states that result from simulated actions.

3. An agent must be able to behave in a way that allows goals to be accomplished.

4. An agent must be able to select actions for internal simulation.

5. An agent must be able to evaluate the sensory state to assess whether or not that action is relevant to the achievement of the goal.

These are necessary and sufficient conditions so that, in principle at least, no other conditions need be met. All five are strongly related to the assumptions made by the simulation hypothesis and the operation of forward and backward models. While each condition is important, the fifth condition turns out to be particularly relevant to a discussion we have yet to have on the subject of representation and we will return to it in the next chapter.

7.5.6   The Basis for Internal Simulation

There is an important unanswered question in our treatment of internal simulation: what is the origin of the internal model that the simulation is based on? In other words, how does the simulation process get started and how does it develop? One possible answer lies in the *inception of simulation hypothesis*[38] which asserts that internal simulations in young infants are formed by re-enacting sensory-motor experiences in dreams. The models that arise from these simulations are validated while awake and subsequently refined. As the child develops, the simulations become more accurate and reliable, adjustments are needed less and less, and internal simulation can be used increasingly in everyday cognitive activities. This inception and adjustment of simulation models is a form of learning and is similar to an idea we met briefly in the previous chapter, Section 6.1.3, Sidenote 22. There, we noted that, in the brain, the hippocampal formation and the neo-cortex may form a complementary system for learning, with the hippocampus facilitating rapid auto- and hetero-associative learning which is later used to reinstate and consolidate learned memories in the neo-cortex. What we didn't mention there is that this reinstatement can occur off-line as well as on-line, *e.g.* in mental rehearsal, recollection, and during sleep.[39]

[38] The inception of simulation hypothesis, according to which dreams in young children play a role in the formation and refinement of internal simulation, was formulated by Serge Thill and Henrik Svensson [344]. It formed the basis of computational experiments using a robotic simulator. These experiments showed that robot "dreams" can lead to faster development of improved internal simulation during waking behaviour [337].

[39] For more details, see the review "Why there are complementary learning systems in the hippocampus and neocortex: insights from the successes and failures of connectionist models of learning and memory" by James McClelland and colleagues [299].

7.5.7   Afference and Efference

The language in internal simulation can be confusing. Two terms
that are used frequently are *afference* and *efference*. Afference
refers to sensory input; efference refers to motor output. Thus,
we speak of afferent signals and efferent signals to refer sensory
stimuli and motor commands respectively. You will also see the
term *efference copy*. This just means that the motor commands are
sometimes directed to other sub-systems, not just the actuators
themselves, be they muscles or motors. The copy could be a literal copy, stored temporarily in memory, or it could also just be
a direct connection via a feedback loop. In internal simulation,
efference copies are typically re-directed back to sub-systems
responsible for sensory interpretation in this manner.

7.5.8   Internal Models for Motor Control

Our discussion of prospection and internal simulation so far
has focussed on anticipation over quite long periods of time:
several seconds or minutes for some behaviours and possibly
much longer. However, internal simulation and internal models
are also involved in short-term prospection less than one second. This is particularly true in the area of motor control and
trajectory planning.[40] The case for prospective control is very
compelling: the delays in acquiring and processing feedback
(i.e. information that captures the error between a desired and
an actual state) in biological system are in the region of 150–250
ms. This is far too long to be effective in controlling the agent's
movement. Again, both forward models and inverse models
come into play. Here, the forward internal models predict both
the future state of the body part being controlled, e.g. the hand
when reaching for some object, on the basis of efference copies
of the motor commands that are being issued. The inverse models determine the motor commands that are required to achieve
some desired state. The predicted or desired state in question
might be position, velocity, acceleration of the body part being
controlled (and related forces) or it might be some set of associated sensory features. These are referred to as a dynamic
model (because they model the system's dynamics) and a sen-

[40] The review article "Internal models in the cerebellum" by Daniel Wolpert
and colleagues [345] sum-
marizes the role of forward
and inverse models in motor
control and discusses the
possibility that the cerebellum uses a model, and
perhaps several models,
of the motor system and
the physical embodiment
in prospectively guiding
an agent's actions. A later
review "Principles of sensorimotor learning" [346]
discusses how these models
are learned.

sory output model (providing, e.g., the predicted afferent consequences of a movement), respectively. Again, multiple models are thought to be involved, all operating simultaneously, just as with prospection and internal simulation over longer periods as discussed above.[41]

7.6 Forgetting

This is a chapter on memory, prospection, and internal simulation, so it is appropriate to close it by discussing briefly one of the most notable characteristics of cognitive agents: they forget. Memory fades in animal brains without repeated stimulation, or re-consolidation as it is sometimes called. On the other hand, for convenience, memory in artificial cognitive systems is often assumed to be persistent.[42] The focus is usually placed on how the contents of memory can be learned effectively, i.e. the emphasis is on what needs to be learned, how best to learn it, and how best to represent it. While the simplification of memory persistence is convenient, it is conceivable that an important aspect of memory and cognition is lost in this simplification. By recognizing the fact that memory traces fade, we are forced into addressing how this fading can be alleviated and what role such a mechanism might play in the overall cognitive process. Unfortunately, the causes for this process of forgetting — and the dual process of memory retention — are not well understood.[43] Furthermore, it appears that different memory retention mechanisms apply in different forms of memory.

In short-term memory, forgetting is an intrinsic property, exactly because short-term memory is a transient phenomenon. As we mentioned above, short term memory depends on persistent elevated firing rates in local sub-populations of neurons in the absence of external cues. Forgetting happens naturally when the activity of these neurons decreases.

In long-term declarative episodic memory, it is less clear what factors influence forgetting. Possible causes include decay, interference between memories, and interference from mental activity and memory formation. It is even less clear what are the mechanisms of forgetting in non-declarative procedural memory. What

[41] Mitsuo Kawato's article "Internal models for motor control and trajectory planning" [347] gives a useful overview of the case for forward and inverse internal models, in general, and of *multiple* paired forward and inverse models, in particular.

[42] Memory in artificial cognitive systems is not always assumed to be persistent. For example, since episodic memory reflects the experiences of a cognitive agent, it can grow continuously. For that reason, forgetting algorithms are sometimes used. Andrew Nuxoll and colleagues compare forgetting algorithms for episodic memory in three episodic memory systems in [348] and conclude that activation-based algorithms can be more effective at selecting the memories with lower utility and, hence, the ones that should be removed.

[43] See John Wixted's article "The Psychology and Neuroscience of Forgetting" [349] for an overview of why and how people forget and a summary of how our understanding of forgetting has evolved over the past 30 years.

is clear though is that long-term memory of itself is not persistent in the normal course of events in a cognitive agent. Where long-term memory forgetting is modelled in artificial cognitive systems, time-dependent decay functions are often used rather than interference between memories and the degree to which memory decays or fades in long-term memory is modelled as a logarithmic, power, or exponential function of time.[44]

[44] The consolidation and decay of long-term memory can be realized using what is referred to as a leaky integrator model (see [350]). John Staddon has shown that cascaded leaky integrators provide a model of decay which is more consistent with Jost's law — a general principle of memory which states that newer memories decay at a faster rate than old memories — than simple exponential decay [351].

8

Knowledge and Representation

8.1   Introduction

Memory and knowledge are intimately related. By focussing
our attention on memory in the previous chapter, we introduced
many of the concepts that are relevant to knowledge. In this
chapter we will build on these and discuss how the principles we
have learned in the previous chapter can be applied directly to
our understanding of knowledge.

Knowledge, arguably, needs to be represented somehow. The
need for representations and the form they take are contentious
issues in cognitive science. We will spend some time discussing
the various issues on either side of the debate. In the process,
we will introduce the *symbol grounding problem*: the problem of
how symbolic knowledge becomes grounded in experience and
acquires meaning for a particular cognitive agent.

Following this, we ask how an agent acquires knowledge and
how it shares it. We already met some of the issues associated
with these questions in Chapter 6 on development and learning,
and we will revisit and expand on them here. The two concerns
— acquiring and sharing knowledge — are clearly related, especially in the context of several agents interacting with one
another. This provides a fitting introduction to the topic of the
next and final chapter of the primer: social cognition.

8.2   The Duality of Memory and Knowledge

In the previous chapter, we referred to knowledge sparingly. As
we explained there, we did this for the specific reason that we
did not want to over-complicate matters too soon. However, the
time has come to face this issue head on, identify the complica-
tions, and explain how they can be avoided by being explicit and
clear about the assumptions we make when discussing knowledge.

   To begin with, we emphasize that there is a clear link between
memory and knowledge and we can qualify different types of
knowledge just in much the same way as we did with memory.
Thus, we differentiate between declarative, procedural, episodic,
and semantic knowledge, depending on the nature of what is
known: knowledge of facts, skill-based knowledge (know-how),
knowledge of specific incidents and events, or knowledge of abstract concepts, respectively. That's the easy part. The difficulty
is the hidden assumptions about the nature of that knowledge.

   As with so many of the complicated topics we have encountered in studying cognitive systems, most of the issues that
cloud the issue of memory and knowledge can be traced to the
paradigm of cognitive science that underpins our assumptions,
be it cognitivist or emergent (and, therefore, embodied to some
extent).

   In the cognitivist paradigm, the situation is very clear. As we
saw in Chapters 2 and 3, knowledge is the content that complements the cognitive architecture. Together, they provide the complete cognitive model. Very often, as a natural consequence of
cognitivism's close relationship with classical AI and the physical symbol system hypothesis in particular (see Section 2.1.2),
knowledge, even procedural knowledge, is assumed to be symbolic. The validity of these two tacit assumptions — knowledge
as content and knowledge as symbols — is a given from the perspective of cognitivism. However, when treating knowledge from
the perspective of the emergent paradigm they are problematic because neither assumption may be valid. In the preceding
sections, we viewed memory as both content and process: as a
mechanism for prediction and recollection. Knowledge, from

this perspective, is no more and no less than the manifestation of that process: it is what emerges when memory works effectively. Viewing knowledge and memory as duals — complementary aspects of the same thing — keeps the process aspect of knowledge in focus without pre-judging the nature of the representation of that knowledge, be it symbolic or non-symbolic (or, as it is sometimes called, sub-symbolic).

8.3   Representation and Anti-representation

In the introduction to this chapter, we said that *arguably* knowledge needs to be represented. It turns out that one of the most hotly debated topics in cognitive science is whether or not cognitive systems use representations and, if they do, what is the nature of these representations. Let's first address the issue of representation *vs.* non-representation[1] and then turn our attention to the various aspects of representation itself.

We first met the representation *vs.* non-representation debate in Chapter 5, Section 5.5, when we discussed the replacement hypothesis on embodiment (see also Chapter 5, Sidenote 20). According to this hypothesis, there is no need for the cognitive system to represent anything because all the information a cognitive system needs is already immediately accessible as a consequence of its non-stop real-time sensorimotor interaction with the world around it. As we noted in Chapter 5, Section 5.5, one of the arguments against this non-representational position is that none of the examples put forward in its favour are "representation hungry"[2] in the sense that they involve situations where the cognitive agent has to act on the basis of knowledge which is *not* presently available to it.

Now, if we admit that knowledge is represented somehow in a cognitive system, there still remain some fine distinctions that derive from the differences between the cognitivist and emergent paradigms concerning what is represented and how it is represented.

[1] See Andy Clark's and Josefa Toribio's article "Doing without Representing?" [206] for an overview of the debate on representation and non-representation in cognitive science and an extended argument against the anti-representational case. Henrik Svensson's and Tom Ziemke's paper "Embodied Representation: What are the Issues?" [352] gives a succinct overview of the debate from the perspective of embodied cognition.

[2] As we noted previously, the idea of "representation hungry" problems is introduced in Andy Clark's and Josefa Toribio's paper "Doing without Representing?" [206], pp. 418–420, and discussed further in Andy Clark's book *Mindware* [33].

8.3.1   Representation and Sharing Knowledge

The cognitivist approach holds that a cognitive system's representation of the world is a direct one-to-one mapping between an internal state (typically, encapsulated in symbolic form) and its counterpart in the real world. This mapping is established by perceptual processes and it assumes that the things we perceive in the world *are* just as we perceive them. Consequently, all other perceptually-equipped cognitive systems perceive the world in the same way. The representation of what exists in the world is therefore a faithful model of the world. Because the real world is not relative to any cognitive observer — *it is what it appears to be* — all properly constructed models must therefore be compatible. As a consequence of this, it is possible for one cognitive agent to assume that its model is entirely consistent with that of another cognitive agent.

As a result, sharing knowledge among cognitive systems poses no problems in principle: all knowledge representations are consistent exactly because they are are derived from a unique and absolute real world. An important consequence of this viewpoint is that it is therefore feasible for a human designer to implant knowledge directly in an artificial cognitive system, since the designer's model is by definition compatible with that of any other cognitive system. This is the basis for the representation and embedding of knowledge in the Soar cognitive architecture, for example (see Section 3.4.1).

The counter-point to this view is the one advocated by the emergent systems community. Whilst acknowledging that all cognitive systems are embedded in, and are situated in, a shared reality, they hold that one can't automatically assume that your perceptions and cognitive understanding of that reality are necessarily identical with all other cognitive agents. On the contrary, the emergent position — and the embodied cognition position — is that our perceptions and understanding are fundamentally linked to the manner in which you interact with that reality (or the manner in which you are structurally coupled with it, to use the terminology we introduced in Chapters 2 and 5). Your perceptions, and therefore your representations of what it is you are perceiving, are shaped by your actions and the range of possible actions you can perform.

From this perspective, any knowledge that might exist in a cognitive system is relative to the situatedness of that system and its history of interaction. There is always the potential to see the world differently because there is always the possibility to alter your space of interaction or structural coupling. An important consequence of this viewpoint is that it is *not* possible for a human designer to implant knowledge directly into an artificial cognitive system, since the designer's model is the result of her or his personal history of interaction with the world and is dependent on her or his space of action, a space that is by definition different from the cognitive system he or she is designing. Knowledge, then, must be acquired by an embodied cognitive agent by learning.

Unfortunately, and setting aside for the sake of convenience the anti-representational stance,[3] the situation in the emergent cognition paradigm, and in embodied cognition in particular, is still somewhat confused because there is no broadly agreed way of identifying what constitutes a representation in the first place.[4]

8.3.2 *What Qualifies as a Representation?*

One might argue that any stable state of a cognitive system, and of its memory in particular, that correlates with events in the world is a representation. This may not be a valid argument, however. Just because stable states of a cognitive system are strongly correlated with events in the world does not, according to some experts in the field, necessarily make them representations. To qualify as a representation these states — these "stand-ins" for the things in the world that are not immediately accessible to the cognitive agent — must also be used for some purpose or function and must be generally available for such use by the cognitive system.[5]

Another way of putting this is to say that a representation must play an active causal role in generating the system's behaviour. By now, you should be able to see how this view of a

[3] Pim Haselager and colleagues deem the debate between representationalists and anti-representationalists to be one that can't be resolved; see their article "Representationalism vs. anti-representationalism: a debate for the sake of appearance" [353].

[4] Again, refer to reference [352] for an overview of the types of mechanisms in embodied cognition that can be considered candidates for representationhood, and the criteria that a potential mechanism must satisfy to be considered a representation (at least from the perspective of embodied cognition).

[5] The additional qualification that for a brain state, e.g., to be considered a representation of something it must not only be correlated with that something but it must also be used by the cognitive system for some purpose or function is highlighted by Lawrence Shapiro in his book *Embodied Cognition* [83] and by Andy Clark in his article "The Dynamical Challenge" [196].

representation mirrors the way we characterized memory systems above as pervasive and active components of the overall cognitive architecture. It is also consistent with internal simulation (or emulation) as a mechanism for achieving the anticipatory prospective capacity that is one of the hallmarks of cognition.

8.3.3   Weak and Strong Representation

We can distinguish between weak and strong representations. Weak representations correspond to events that are currently accessible by our senses while strong representation correspond to those that are not (e.g. objects that are out of sight or that we saw previously).[6] Strong representations are required in circumstances where the events to be represented might no longer be present, might not even exist, or might be counter-factual: the opposite of affairs as they appear to be.[7] These situations require strong internal representations, typically to allow the cognitive agent to function prospectively.

There is also a form of "representation" which falls somewhere between representationalism and anti-representationalism. It derives from the enactive approach to emergent cognitive systems we discussed in Chapter 2, Section 2.2.3.

Recall that an enactive system constructs its own model of the world as a consequence of the mutual structural coupling between the agent and the world in which it is situated and embedded. We referred to this process as *sense-making* and we noted that the resultant knowledge says nothing at all about what is really out there in the environment. It doesn't have to: all it has to do is make sense for the continued existence and autonomy of the cognitive system. This self-generated knowledge cannot be said to be a "re-presentation" of the agent's world — for the principled reason that an enactive system is organizationally-closed[8] — but nevertheless the internal states do play a "stand-in" part in the agent's cognitive processes, its internal simulation, and prospection. Thus, they are representations in an alternative sense that doesn't imply a re-presentation of anything external, but is instead a re-presentation of an agent's self-constructed

[6] The distinction between *weak representations* and *strong representations* is explained by Andy Clark in his article "The Dynamical Challenge" [196]. Lawrence Shapiro quotes him in his book on embodied cognition [83]: "Weak internal representations are those ... that merely have the function of carrying information about some object that is in contact with sensory organs. Unlike a map or a picture, these representations persist for only as long as the link between them and the world remains unbroken. ... These weak representations are ideal for 'inner systems that operate only so as to control immediate environmental interactions' [23], p. 464."

[7] The situations in which *strong internal representations* are required typically arise in problems that Andy Clark and Josefa Toribio refer to as being *representation hungry*: the problems involve reasoning about absent, non-existent, or counterfactual events [206].

[8] See Section 4.3.6 and Side-note 32 for an explanation of *organizational closure* and the related concept of *operational closure*.

understanding of its world derived from its experience as an autonomous organizationally-closed entity, a construction that plays a part in achieving adaptive anticipatory behaviour.

8.3.4   Radical Constructivism

The constructive aspect of enactivism is referred to as *constructivism*[9] and, occasionally, *radical constructivism*.[10] The qualification *radical* is applied to constructivism to emphasize that the principles of constructivism have to be applied at every level we chose to describe a cognitive system. Strictly-speaking, (radical) constructivism rejects representationalism, but only in the sense that representationalism assumes an external world — a reality — to which cognitive agents have direct access and can represent. Constructivism does allow for knowledge; it simply stipulates that knowledge is the result of an active process of construction whereby the cognitive agent determines through its structural coupling with its environment what matters for its survival and what doesn't.[11] This is "sense-making," in the language of enaction, or model generation, in the language of computational modelling.

Before finishing up on this topic, a word on symbolic representations is in order. While explicit symbolic representations are the life-blood of cognitivist systems and there other implicit non-symbolic forms of representation, symbolic encoding of knowledge is only strictly necessary when you need to encapsulate that knowledge in some linguistic form to effect its representation and communication externally to another agent. This linguistic communication may be written, graphic, or spoken. Thus, according to some experts, symbolic knowledge representations may exist as mechanisms of communication of meaning rather than being the mechanisms of the cognitive processes which gave rise to them: "The point of having a symbolic representation of knowledge is exactly its utility and ability to convey and share meaning between cognitive agents."[12]

[9] For an introduction to *constructivism*, see Alexander Riegler's editorial in the inaugural issue of *Constructivist Foundations* [354].

[10] The definitive work on radical constructivism is Ernst von Glaserfeld's book *Radical Constructivism* [46]. He has also published a short overview of the field in a paper entitled "Aspects of radical constructivism" [355].

[11] The two basic principles of radical constructivism are [355]: (1) Knowledge is not passively received either through the senses or by way of communication, but it is actively built up by the cognizing subject; (2) The function of cognition is adaptive and serves the subject's organization of the experiential world, not the discovery of an objective ontological reality.

[12] This quotation is taken from a book by George Lakoff and Rafael Núñez on the origin of mathematics, perhaps the most refined manifestation of symbolic communication, in which they argue that mathematics results from human cognition rather than having some transcendent Platonistic existence. This does not take from its utility or meaningfulness: "[Mathematics] is precise, consistent, stable across time and human communities, symbolizable, calculable, generalizable, universally available, consistent within each of its subject matters, and effective as a general tool for description, explanation, and prediction in a vast number of everyday activities" [356].

8.4   The Symbol Grounding Problem

If a cognitive system has some form of symbolic representation of the world around it — some set of tokens that denote objects in the agent's world — the question arises as to how the representation acquires meaning? How do purely symbolic representations acquire semantic content? This might seem like an innocent question but it is made difficult by the fact that symbols systems, which we described in Section 2.1.2, are governed by purely syntactic processes. That is, the atomic symbols, the strings of symbols, and the symbol-based rules that define the manipulation and recombination of symbols and strings of symbols are all defined in terms that make no reference to what these symbols mean. On the other hand, they are all "semantically interpretable": that is, the syntax can be assigned a semantic meaning so that symbols and strings of symbols can represent objects, events, or concepts, and describe them or stand in for them. The problem is how to assign this meaning. This is the symbol grounding problem.[13] The key idea is that symbolic representations have to be grounded bottom-up in non-symbolic representations of two kinds:

1. Iconic representations, which are derived directly from sensory data; and

2. Categorical representations, based on the output of both learned and innate processes that detect invariant features of object and event categories from these sensory data.

Higher-order symbolic representations can then be derived from these elementary symbols. The iconic representations allow you to discriminate between different objects and the categorical representations allow you to identify an object encapsulated by its iconic representation as belonging to a particular class or category of objects (this is why the features have to be invariant: they don't change significantly within a given category). Both types of representation, iconic and categorical, are non-symbolic and, therefore, a non-symbolic process is required to learn the invariances and thereby form the categories. Usually, we resort to some form of connectionist approach (see Chapter 2, Section

[13] The symbol grounding problem was introduced by Stevan Harnad in a classic paper [40] in 1990. A restricted form is the *anchoring problem* [357] which differs from the more general symbol grounding problem in a number of ways. It is concerned only with artificial systems and focusses on establishing a relationship between a symbolic label denoting some object and the sensory perception of that object, maintaining that relationship over extended periods of time, even when that object cannot be seen. Also, it is concerned only with grounding physical objects and doesn't address the grounding of abstract concepts such as war or peace.

2.2.1) to accomplish this mapping and form the categorical representation. As a consequence, according to this argument, a grounded symbol system is a *hybrid system*: a combination of symbolic and emergent approaches. We introduced hybrid systems in Chapter 2, Section 2.3, and gave an example of a hybrid cognitive architecture in Chapter 3, Section 3.4.3.

Not everyone agrees with this characterization of the symbol grounding problem. An alternative viewpoint is that internal symbolic representations are the result of ontogenetic development and that they are are *tethered* to the world through sensory perception rather than being *grounded*.[14] The distinction is an important one. Symbol grounding implies that the meaning of a symbol is derived bottom-up by abstraction from direct sensory experience. The need for symbol grounding in this sense is a direct consequence of adopting a cognitivist approach to cognition. Symbol tethering is quite different. It arises through a rich process of structural coupling with the world. With symbol tethering, the symbols don't derive directly from the sensory data, they derive from development, the process of developing new items of knowledge that are specific to the embodiment of the agent in question. Thus, symbol grounding is required only if one adopts a cognitivist approach; symbol tethering is more neutral in the sense that it makes no strong claims about the relationship between world and respresentation, or the necessary uniqueness of these representations.

The difference between symbol grounding and symbol tethering is very much the same as the difference we discussed in Chapter 2, Section 2.2.2, on the difference between representations that *denote* objects and those that *connote* objects.

[14] The concept of *symbol tethering* was introduced by Aaron Sloman [301]. His seminar on symbol grounding and symbol tethering [358] is a good place to start reading about the difference between the two ideas. Symbol tethering is sometimes referred to as *symbol attachment*.

8.5 Joint Perceptuo-motor Representations

In Section 7.5 above we remarked on the fact that mental imagery, viewed as another way of expressing the process of internal simulation, comprises both visual imagery (or, better still, perceptual imagery) and motor imagery. More importantly, though, we noted that these two forms of imagery are tightly entwined: they complement each other and the simulation of perception and covert action both involve elements of visual and motor imagery. We subsequently discussed the idea that the classical separation of declarative and procedural memory, and of episodic and procedural memory in particular, is being eroded to reflect contemporary understanding of the interdependence of perception and action. This gave us the opportunity to mention the idea of joint perceptuo-motor representations: representations that bring together the motoric and sensory aspects of experience in one framework. We left that thread of our narrative hanging at that point while we discussed the difficult issue of representation in its own right. Having done that, we now come back to address joint perceptuo-motor representations in more detail. We consider here two different approaches: the *Theory of Event Coding* and *Object-Action Complexes*. To set the scene for them, we must first explain the difference between sensory-motor theory and ideo-motor theory.

8.5.1   *Sensory-motor Theory and Ideo-motor Theory*

Broadly speaking, there are the two distinct approaches to planning actions: *sensory-motor* action planning and *ideo-motor* action planning.[15] Sensory-motor action planning treats actions as reactive responses to sensory stimuli and assumes that perception and action use distinct and separate representational frameworks. The sensory-motor view builds on the classic unidirectional data-driven information-processing approach to perception, proceeding stage by stage from stimulus to percept and then to response. It is unidirectional in that it doesn't allow the results of later processing to influence earlier processing. In particular, it doesn't allow the resultant (or intended) action to impact on the related sensory perception.

   Ideo-motor action planning, on the other hand, treats action as the result of internally-generated goals. It is the idea of achieving some action outcome, rather than some external stimulus, that is at the core of how cognitive agents behave. This reflects the view of action described in Chapter 6, Section 6.3, with action being initiated by a motivated subject, defined by goals, and guided by prospection. The key point of the ideo-motor principle is that

[15] For a good overview of ideo-motor theory, read Armin Stock's and Claudia Stock's "A short history of ideo-motor action" [359]. Sensory-motor and ideo-motor models are sometimes written sensorimotor and ideomotor models; either formulation is correct.

the selection and control of a particular goal-directed movement depends on the anticipation of the sensory consequence of accomplishing the intended action: the agent images (e.g. through internal simulation) the desired outcome and selects the appropriate actions in order to achieve it.

There is an important difference, though, between the concrete movements comprising an action and the higher-order goals of an action. Typically, actors do not voluntarily pre-select the exact movements required to achieve a desired goal. Instead, they select prospectively-guided intention-directed goal-focussed action, with the specific movements being adaptively controlled as the action is executed. Thus, ideo-motor theory should be viewed both as an anticipatory idea-centred way of selecting actions and as a way of bridging the higher-order conceptual representations of intentions and goals[16] with the concrete adaptive control of movements when executing that action.[17]

In contrast to sensory-motor models, ideo-motor theory assumes that perception and action share a common representational framework. Because ideo-motor models focus on goals, and because they use a common joint representation that embraces both perception and action, they provide an intuitive explanation of why cognitive agents, humans in particular, are so adept at and predisposed to imitation.[18] The essential idea is that when I see somebody else's actions (and remember: actions are goal-directed) and the consequences of these actions, the representations of my own actions that would produce the same consequences are activated.

At first glance, ideo-motor theory seems to present a puzzle: how can the goal, achieved through action, cause the action in the first place? In other words, how can the later outcome affect the earlier action? This seems to be a case of *backward causation*, i.e. causation backwards in time.[19] The solution to the puzzle is prospection. It is the anticipated goal state, not the achieved goal state, that impacts on the associated planned action. Goal-directed action, then, is a centre-piece of ideo-motor theory, which is also referred to as the *goal trigger hypothesis*.[20]

[16] Michael Tomasello and colleagues note that the distinction between intentions and goals is not always clearly made. Taking their lead from Michael Bratman [360], they define an intention as a plan of action an agent chooses and commits itself to in pursuit of a goal. An intention therefore includes both a means (i.e. an action plan) as well as a goal [361].

[17] See "Hierarchy of idea-guided action and perception-guided movement" by Sasha Ondobaka and Harold Bekkering [362] for a review of the ideo-motor principle in the context of higher-order conceptual goals.

[18] See Marco Iacoboni's article "Imitation, Empathy, and Mirror Neurons" [341] for an explanation of the links between ideo-motor theory, imitation, and the mirror neuron system. Also refer back to the earlier discussion of imitation and mirror neurons in Chapter 6, Section 6.1.2, and Chapter 5, 5.6, respectively.

[19] *Backward causation* — later events influencing earlier events — is not the same as the *downward causation* we met in Chapter 4, Section 4.3.5, where global system activity influences the local activity of the system's components.

[20] Bernhard Hommel and colleagues refer to ideo-motor theory as the "goal trigger hypothesis" in their article on the Theory of Event Coding [342].

8.5.2   The Theory of Event Coding

The Theory of Event Coding (TEC)[21] is a representational framework for combining perception and action planning. It focusses mainly on the later stages of perception and the earlier phases of action. As such, it concerns itself with perceptual features but not with how those features are extracted or computed. Similarly, it concerns itself with preparing actions — action planning — but not with the final execution of those actions and the adaptive control of various parts of the agent's body. The main idea is that perception, attention, intention, and action all work with a common representation and, furthermore, that action depends on both external and internal causes.

TEC is intended to provide a basis for combining both sensory-motor and ideo-motor action planning: to be a joint representation that serves both sensory-stimulated action and prospective goal-directed action. The core concept in TEC is the *event code*. This is effectively a structured aggregation of distal features of an event in the agent's world, i.e. features that the agent observes from some distance. In TEC, these are called *feature codes*. They can be relatively simple (e.g. colour, shape, moving to the left, falling) or more complex, such as an affordance.[22] Also, TEC feature codes can emerge through the agent's experience; they don't have to be pre-specified. Remember, TEC is a framework and doesn't concern itself with how these features are made explicit or computed.

A given TEC feature code (of which there are many) is associated with both the sensory system and the motor system (see Figure 8.1). Typically, a feature code is derived from several proximal sensory sources (sensory codes) and it contributes to several proximal motor actuators (motor codes). They are proximal because they are part of the agent's body, as opposed to being distal and disconnected from the agent's body. Thus, sensory and motor codes capture proximal information whereas feature codes (however they are computed) in the common joint TEC representation capture distal information about events in the world.

Each *event code* comprises several feature codes representing

[21] The Theory of Event Coding (TEC) was proposed by Bernhard Hommel and colleagues in 2001 [342]. As well as setting out the main tenets of the theory, their paper also provides a good overview of the two distinct approaches to planning actions: sensory-motor action planning and ideo-motor action planning.

[22] To refresh your memory of affordances, refer back to the second last paragraph of Chapter 5, Section 5.6, and to Sidenote 52 in Chapter 2.

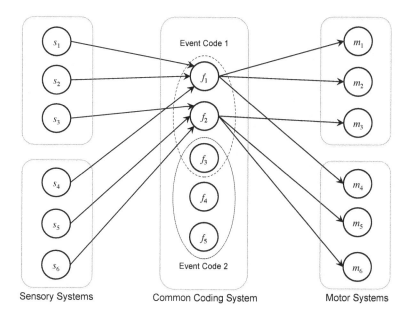

Figure 8.1: Features in the Theory of Event Coding combine sensory information from a variety of sensory modalities. In this example, sensory information s_1, s_2, s_3 and s_4, s_5, s_6 comes from two sensory systems and converges onto two abstract feature codes f_1 and f_2 in a common-coding system. These features spread their activation to codes belonging to two different motor systems: m_1, m_2, m_3 and m_4, m_5, m_6, respectively. Feature codes, which can be activated by external stimulation or internal processes, are integrated into separate event codes. Features, e.g. f_1, can be shared between event codes. Adapted from [342].

some event, be it a perceived event or a planned event. Feature codes associated with an event are activated both when the event is perceived and when it is planned. Of course, because features can be elements of many event codes, the activation of a given feature effectively primes, i.e. predisposes, all the other events of which this feature is a component. Activation of features alone, though, is not the end of the story. The features that make up an event are bound together: integrated into some event code. The nature of the binding isn't specified in TEC but the effect of binding is a form of event code suppression in which one event will interfere with and inhibit other events that share some of the dominant event codes features.

8.5.3 Object-Action Complexes

Our second example of a joint representation of action and perception comes from the robotics domain. The representation is referred to as an Object-Action Complex, or OAC for short (pronounced "oak").[23] An OAC is a triple, i.e. a unit with three components: (E, T, M), where

[23] Norbert Krüger and colleagues explain the way that Object Action Complexes (OACs) provide symbolic representations of sensory-motor experience and predictive behaviours in their article "Object-Action Complexes: Grounded abstractions of sensory-motor processes" [363]. This article also provides some examples of how OACs have been used by an artificial cognitive system, a humanoid robot in this case, to learn the affordances of simple objects.

- E is an "execution specification"; think of it as an action.

- $T : S \rightarrow S$ is a function that predicts how the attributes
  that characterize the current state of the agent's world will
  change if the execution specification is executed. Think of T
  as a prediction of how the agent's perceptions will change as
  a result of carrying out the actions given by E. S is just the
  space of all possible perceptions of the agent.

- M is a statistical measure of the success of the OAC's past
  predictions.

Thus, an OAC combines the essential elements of a joint representation — perception and action — with a predictor that links
current perceived states and future predicted perceived states
that would result from carrying out that action. To a large extent, an OAC models an agent's interaction with the world as
it executes some motor program (this is referred to a low-level
control program CP in the OAC literature). For example, an
OAC might encode how to grasp a object or push an object into
a given position and orientation (usually referred to as the object pose). OACs can be learned and executed, and they can be
combined into more complex representations of actions and their
perceptual consequences.

8.6   Acquiring and Sharing Knowledge

In Section 8.3 above we highlighted an important difference
between the cognitivist and the emergent approaches to cognition regarding their treatment of knowledge: that it is possible
in principle to directly share knowledge between cognitivist
systems, whereas it is not possible in principle to do so for emergent ones. Instead, an emergent system has to learn its knowledge. A cognitivist system can learn knowledge too, of course:
the issue is the possibility or not of direct sharing. In this section,
we look at how cognitivist agents — robots in particular — share
knowledge directly with humans and other robots. To complete
the picture, we also look at how emergent agents — again, cognitive robots — acquire knowledge and know-how by observing

other agents, i.e. by learning from demonstration. This approach to indirect sharing of knowledge and know-how among cognitive agents is related to the topic of imitation which we discussed in Chapter 6, Section 6.1.2.

8.6.1 Direct Knowledge Transfer — the Cognitivist Perspective

One the greatest strengths of the cognitivist paradigm, and at the same time one of its potential weaknesses, is the assumption that all cognitive agents have access to the same understanding of the world around them. The strength of this position is that it means that the knowledge possessed by one agent is inherently compatible with the understanding mechanisms — the faculties for perception, reasoning, and communication — of another agent. This strength has been widely exploited by all cognitivist systems to allow humans to embed representations of their knowledge in cognitive systems, very often as symbol-based rules as we saw in Chapter 3, Section 3.1.1 on cognitivist cognitive architectures. The weakness lies in the fact that a programmer's knowledge is limited to what he or she knows and considers important enough to encode and embed in the cognitive agent. Of course, a cognitivist cognitive agent can also learn for itself, but its understanding — its reasoning capability — is often bounded by the programmer's initial knowledge.

Recently, however, there have been developments designed to overcome this weakness by allowing robots to share their knowledge and their experience, and autonomously mine other sources of knowledge. Thus, as one robot learns how to solve a given task, it can make that knowledge available to other robots. This is the rationale for RoboEarth, [24] a fast-growing world-wide open-source framework that allows robots to generate, share, and reuse knowledge and data. The RoboEarth architecture provides for various types of knowledge: global world models of objects, environments, actions, all linked to semantic information. Different representations are used: for example, images, point clouds, and 3-D models for objects, maps and coordinates for environments, and human-readable action recipes for actions and task descriptions. This information is linked with a

[24] The *RoboEarth* framework is described in general terms in an article "RoboEarth: A World-wide Web for Robots" by Markus Waibel, Michael Beetz, and others [34]. A later paper by Moritz Tenorth, Michael Beetz, and colleagues, "Representation and Exchange of Knowledge about Actions, Objects, and Environments in the RoboEarth Framework," provides more technical detail on the various knowledge representations involved and the associated knowledge reasoning techniques use to draw inferences about that knowledge [364].

graph-based semantic representation that allows the robot to understand how to use all the various knowledge components.

RoboEarth is effectively a world wide web for robots: a web-based resource that robots can use to exchange knowledge among each other and benefit from the experience of other robots, customizing that knowledge to suit their own particular circumstances. It is this ability to customize that particularly distinguishes RoboEarth. It is not just a repository of object, environment, and task data: it includes also the semantic knowledge that encodes the meaning of the content in terms of the relationships between the various entities in a way that allows the robot to decide if an object model will be useful in some given task, for example.

The core of the framework is a language that allows all the relevant information to be encoded, exchanged, and re-used by the robots. For example, it provides ways of describing actions and their parameters, objects and their positions and orientations (poses), and models for recognizing objects. It also provides a way of exchanging the meta-information — information about the information — such as coordinate reference frames and units of measure. Since the robots that are hooked up to RoboEarth will differ significantly in size, shape, and sensory-motor capabilities, the language also allows models of the robot's configuration to be captured as well as specifications for prerequisite components that a robot must have to make use of each piece of information. Finally, it also provides a way to match these specifications to the robot's capabilities so that it can check to see if it is missing any components.

While a world wide web exclusively for robots has clear advantages, there is a lot of information intended for human use on the normal world wide web that could be very useful for robots too.[25] Cognitive robots could use sites that provide step-by-step instruction on how to do things to formulate a plan for undertaking a task. Shopping sites and image databases provide pictures that can be used by the robot to search its own environment for objects. Other sites provide access to 3-D CAD models of household items that can be used to plan grasping actions. Encyclopedic knowledge-bases provide information on the rela-

[25] The article "Web-enabled Robots" by Moritz Tenorth, Michael Beetz, and colleagues describes a number of ways in which the world wide web is being used to provide knowledge for robots to help them carry out everyday tasks [365].

tionships between different objects.[26] Common-sense knowledge is rarely formalized but this too is also available As an example, a home-help robot might need to use milk in preparing a meal. The information it acquires about milk from a shopping site will provide the appearance of a milk carton and it will also reveal that it is perishable. Common sense knowledge would provide the information that perishable goods are stored in the refrigerator and encyclopedic knowledge provides information about refrigerators. By combining all this information, the robot can determine where to look to find the milk. The challenge is to provide cognitive robots with the ability to access and combine all this knowledge, as required, when planning a given task.[27]

Finally, some terminology: the sharing of knowedge, both robot-oriented and human-oriented, by robots on the internet is sometimes referred to as *Cloud Robotics*.[28]

8.6.2   *Learning from Demonstration — the Emergent Perspective*

As we have noted on several occasions, we do not have direct access to the knowledge of a cognitive system that adheres to the emergent paradigm of cognitive science. Such knowledge is specific to that particular agent and it is acquired through development and learning. We discussed the importance of exploration and social interaction in Section 6.1.1 on development and in Section 6.1.2 we highlighted the natural predisposition that humans have for imitation. In Section 6.1.3 we remarked that imitative learning provides a way of actively teaching a cognitive robot, instructing it on how to achieve a certain action or task. This approach to imparting knowledge to a cognitive robot, particularly know-how in the form of procedural and episodic knowledge, is referred as *Learning from Demonstration* or *Programming by Demonstration*.[29] We will develop this theme a little more here.

Before beginning, it is important to be clear that learning from demonstration is relevant to cognitive robotics in both the cognitivist and the emergent traditions. However, there are four variants of the general approach and just one of these is applicable to emergent cognitive robotics.

[26] Encyclopedic knowledge is often represented by an *ontology*, a specification of knowledge in terms of concepts, their types, properties, and inter-relationships. By the way, this is how ontology is defined in computer science; it has a different meaning in philosophy where it refers to the study of the nature of existence and reality.

[27] A common knowledge representation helps in tackling the problem of combining information from several sources. The *KnowRob* knowledge base is one example of this type of common representation; see Moritz Tenorth's and Michael Beetz's article "KnowRob — Knowledge processing for autonomous personal robots" [366].

[28] The term *Cloud Robotics* was introduced by James Kuffner at the IEEE Conference on Humanoid Robotics in 2010 [367]. A techncial report by Ken Goldberg and Barry Kehoe "Cloud Robotics and Automation: A Survey of Related Work" provides a succinct overview of the area [368].

[29] For a comprehensive overview of the different approaches to learning from demonstration, read the survey by Brenna Argall and colleagues [300]. The article by Aude Billard and colleagues in the *Springer Handbook of Robotics* provides a good overview of robot programming by demonstration [369], as does the article by Rüdiger Dillman and colleagues [370].

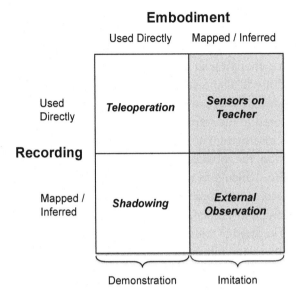

Embodiment

Used Directly   Mapped / Inferred

	Used Directly	Mapped / Inferred
Used Directly	*Teleoperation*	*Sensors on Teacher*
Mapped / Inferred	*Shadowing*	*External Observation*

Recording

Demonstration   Imitation

Figure 8.2: The four types of learning from demonstration. In the top row, the required teaching data is recorded and can be used directly by the robot; in the bottom row, the required data is available indirectly and so has to be inferred by the robot. In the left column — Demonstration — the teaching data matches the robot's body and can be used directly. In the right column — Imitation — the teacher's body doesn't match the robot's body and therefore the teaching data has to be mapped to the robot's body. From [300], © 2009, with permission from Elsevier.

First, let us state exactly what learning from demonstration tries to achieve and then we will see how different approaches to solving this problems yield the four different variants.[30] A cognitive robot needs to know how to act in any given situation: what action is appropriate given the current state of its world. Thus, the goal of the exercise is to learn a mapping from world state to action. This mapping is called a *policy*; here we will refer to it as an *action policy* to remind us that the policy determines appropriate actions based on what's going on in the robot's world.

To learn this action policy, a teacher provides a series of examples — demonstrations — of the world state and the associated actions. These world states typically reflect some situation that requires an action: there is a dirty garment in the laundry basket, so the action might be to put it in the washing machine. As we mentioned previously in Chapter 6, Section 6.1.3, learning from demonstration is a particular form of supervised learning. Often, the world state isn't directly accessible to the robot, so it has to depend on its observations. The learned action policy then tells the robot what action to take based on its observations.

[30] The treatment of programming by demonstration, and the four different categories of approach, follow closely the approach introduced by Brenna Argall and colleagues in their survey [300].

During learning, knowledge is transferred to the robot. That transfer may be explicit or implicit, depending on whether the robot has direct access to the teacher's knowledge — the motion and world state information that characterize the actions being taught — or whether it has to infer it. There are two steps to this transfer: (a) recording of the teacher's actions and (b) determining the correspondence of this data with the robot's embodiment. Either steps can be direct or indirect. If it is direct, then the data can be used by the robot without modification. If it is indirect, the data has to be mapped to the robot, i.e. the information required by the robot to reproduce the action has to be inferred. Since there are two stages, each with two options — directly used and indirectly mapped — there are four different types of learning from demonstration; see Figure 8.2.

The first type is *teleoperation*. Here, a teacher directly operates the robot learner and the robot's sensors records the required actions. Thus, the recording is direct. Since, the teacher and learner embodiments correspond exactly (they are in fact the same body, i.e. the robot), no further mapping is required.

The second type is called *shadowing*. Here, the robot doesn't have direct access to the teachers action data. Instead, it attempts to mimic teachers actions (e.g. by tracking the teacher's action) and it records its own movements through its own sensors. Thus, the recording is indirect. However, the data recorded corresponds directly to the robot's own embodiment.

The third type is referred to as *sensors on teacher*. Here the teacher's movements are recorded directly on the teacher and the learning robot has direct access to this data. Thus, the recording is explicit and direct. On the other hand, the embodiment of teacher and learner robot are different so this data has to be used indirectly and mapped to the particular embodiment of the learner robot.

Finally, the fourth type is called *external observation*. In this case, both the recording and the embodiment correspondence have to be inferred. The robot observes the teacher with sensors that are external to the teacher (in contrast to the *sensor on teacher* type). These could be on the robot itself or arranged around the robot in a sensorized environment. Furthermore, the teacher and robot embodiments differ, such as in the case of a human teaching a 53 degree of freedom humanoid, so that the observed data, i.e. the teacher's movements, has to be mapped to the robot's frame of reference.

The term *demonstration* is used to refer to the first two types: teleoperation and shadowing. *Imitation* refers to either the third or fourth type: sensor on teacher or external observation. However, in the context of our discussion of emergent cognitive systems, only the fourth type — external observation imitation — is applicable because direct knowledge transfer, in either steps of the knowledge transfer process, is not possible.

One final point is relevant. It concerns the manner in which the demonstration — the pairs of world states or observations of world states and associated actions — is presented to the learning robot. There are two options, one is *batch learning* in which the action policy is learned after all the data has been presented. The second is *interactive learning* so that the action policy is updated as the training data is progressively presented to the learning robot. Interactive learning provides the opportunity for the teacher to correct the learned action policy or to suggest ways of improving it, i.e. the teacher not only demonstrates examples of the required action policy but he or she also coaches the robot, fine-tuning its behaviour.[31]

[31] For an example of robot coaching, see the work of Marcia Riley and colleagues who use an interactive learning from demonstration methodology to coach humanoid robots [371]. The key features of their approach is the use of a verbal coaching vocabulary, the ability to refine specific parts of a behaviour, and the ability to clarify instructions or resolve ambiguities through dialogue.

9
Social Cognition

9.1 *Introduction*

So far in this book, cognition has been portrayed as an agent-centred and relatively lonely activity: cognitive systems explore and interact with the world around them but the focus of this interaction has been on structural coupling, with an emphasis on adaptation and autonomy rather than on collective social activity. However, as we noted in Chapter 5 on embodiment and Chapter 6 on development, interaction can go well beyond being embedded in the environment, with other cognitive agents playing a major role in the cognitive activities of an individual cognitive agent. In this final chapter, we place cognitive systems in a collective setting and discuss how cognition in an individual agent takes place in a social milieu. This positions social cognition differently from extended cognition and distributed cognition, as defined in Chapter 5, Section 5.10. Thus, in this chapter we aren't considering how cognition in an individual might directly involve the agent's environment (extended cognition), nor yet are we considering collective cognition in the sense of a group of individual agents collectively exhibiting a capacity for cognition (distributed cognition). Instead, we focus here on how an individual cognitive agent develops and interacts in a social setting involving other cognitive agents.[1]

To provide the basis for discussing the social interaction between two or more cognitive agents, we need to characterize the essence of individual cognition: how a single agent inter-

[1] In this chapter, we restrict our focus to direct interaction between cognitive agents. For a discussion of social cognition where the agents interact indirectly, i.e. though artefacts, using the principles of stigmergy (by which social insects such as ants achieve collective coordination), see the article by Tarja Susi and Tom Ziemke [372].

acts with its environment. At this point of the book, we are now familiar with the idea that cognition focusses on action, and especially on prospective goal-directed action. Action and goals are two of the essential characteristics of cognitive interaction. There are two others. The first of these — intention — captures the prospective aspect of action and goals. Recall the distinction we drew between goals and intentions in Chapter 8, Sidenote 16: an intention is a plan of action an agent chooses and commits itself to in pursuit of a goal. An intention therefore includes both a goal and the means of achieving it. Thus, an agent may have a goal for some state of affairs to exist and an intention to do something specific in pursuit of that state of affairs. Intentions integrate, in a prospective framework, actions and goals. Finally, we have perception: the essential sensory aspect of cognition. However, in the context of cognitive interaction, perception is directed. It is focussed on goals, influenced by expectations: it is attentive.

Although we have mentioned attention on several occasions in the book, we have never actually said exactly what we mean by it. So, before continuing and because it is a key characteristic of cognitive interaction, we will take some time to do so.

Perhaps the closest we got to explaining the attention was in Chapter 5, Section 5.6, where we discussed spatial attention and selective attention (more or less: where we direct our gaze and what sort of things are most apparent to our gaze). We can make this a little more explicit by defining attention as "the temporally-extended processes whereby an agent concentrates on some features of the environment to the (relative) exclusion of others."[2] In closing Chapter 6 we remarked that a cognitive system needs an attentional system which fixates on the goals of actions. This mirrors an important aspect of attention: it is strongly linked to intentionality. It could even be characterized as *intentionally-directed perception*: we focus our attention on what matters to us in pursuit of our goals.[3]

To sum up, then, we characterize the interaction of an individual cognitive agent by the four characteristics of action, goals (or commitment), intention, and attention (understood in the sense of intention-guided perception), all of which have an ele-

[2] This definition of attention is taken from "The Challenges of Joint Attention" by Frédéric Kaplan and Verena Hafner [263], p. 138. However, we must be careful to recognize that attention can be very difficult to pin down and good definitions are very scarce. The title of Chapter 1 of John Tsotsos's book *A Computational Perspective on Visual Attention* [373] and the title of the first section sum this predicament up neatly: "Attention — We All Know What It Is"; "But Do We Really?"

[3] The selectivity of attention was recognized early on by William James who described it as follows: "Everyone knows what attention is. It is the taking possession by the mind, in clear and vivid form, of one out of what seem several simultaneously possible objects or trains of thought" [60]. John Tsotsos used the first of these two sentences as the basis for setting up his discussion of the surprising difficulty in defining attention [373].

ment of prospection. Our goal in this chapter is to show what is
necessary to transform this characterization to one that is rep-
resentative of social cognition: the interaction between two (or
more) cognitive agents. This will involve the notions of joint action, shared goals, shared intentions, and joint attention. As we
will see, this transformation, and these four notions, go far beyond a simple superposition of the notions of individual action,
goals, intentionality, and attention from which they derive. Much
more is involved in social cognition. But we are getting a little
ahead of ourselves and we need to cover some preparatory material to get ready to discuss these issues. To set the scene for this,
we begin with a brief overview of social cognition to introduce
the key elements of social interaction, in general, and helping
other agents, in particular.

9.2   Social Interaction

Social cognition embraces a wide range of topics.[4] The abilities
required for successful social interaction range from reading
faces, detecting eye gaze, recognizing emotional expressions,
perceiving biological motion, paying joint attention, detecting
goal-directed actions, discerning agency, imitation, deception,
and empathy, among many others. To help us navigate our way
through the many related topics, it may help to know where we
are headed. Our target in this chapter is the social capacity of
cognitive agents to collaborate with one another — to engage
in joint action — and we will focus in particular on the manner
in which humans develop the mechanisms they need for collaboration. Along the way, we will explore several related issues,
including the ability to infer the intentions of other agents, to
help them, to share goals and intentions, and ultimately to carry
out joint actions by paying joint attention to the task at hand.

Social cognition — effective social interaction with other cognitive agents — depends on an agent's ability to interpret a wide
variety of visual data that conveys information about the activities and intentions of other agents. As we saw noted in Table 6.1,
newborns have an innate attraction to biological motion and it
has been shown that the ability to process biological motion is a

[4] A research review of
Social Cognition written
by Uta Frith and Sarah-
Jayne Blakemore for the
Foresight Cognitive Systems
Project [374] provides a very
accessible overview of the
field, its scope, and its many
component mechanisms
and processes, as well as
discussing how it is affected
by pathology, e.g. autism
spectrum disorder (ASD).

hallmark of social cognition, providing a cognitive agent with a capacity for adaptive social behaviour and nonverbal communication.[5] The clearest example of this is the ability to read body language, the subtle body movements, gestures, and actions that are an essential aspect of successful interaction between cognitive agents (we assume here that we are dealing with embodied cognitive systems in the sense discussed in Chapter 5). For an agent to interact socially with another cognitive agent, it must be aware of the cognitive state of that agent and be sensitive to changes. In Section 5.7 we discussed the link between the state of an agent's body and its cognitive and affective state, especially during social interaction. Recall that there are four aspects to this link:

1. When an agent perceives a social stimulus, this perception produces bodily states in the perceiving agent.

2. The perception of bodily states in other agents frequently evokes a tendency to mimic those states.

3. The agent's own body states trigger affective states in the agent.

4. The efficiency of an agent's physical and cognitive performance is strongly affected by the compatiblity between its bodily states and its cognitive states.[6]

Because of the strong link between bodily states and cognitive and affective states, the posture, movements, and actions of an agent convey a great deal about its cognitive and affective disposition as well as influencing how another agent will behave towards it.

While social cognition is ultimately about mutual interaction, this interaction doesn't have to be entirely symmetric: for example, one agent can assist another or both agents can assist each other. In the following, we will refer to these behaviours as helping (sometimes adding the qualification *instrumental* helping) and collaboration, respectively. For the moment, it is just important to realize that one (collaboration) develops from the other (instrumental helping).

[5] Marina Pavlova's article "Biological Motion Processing as a Hallmark of Social Cognition" [375] provides a detailed analysis of the tight relationship between the perception of biological motion and social cognitive abilities, noting that individuals who exhibit a deficit in visual processing of biological motion are also compromised in social perception.

[6] Refer again to the article "Social Embodiment" by Lawrence Barsalou and co-workers [207]; it provides many examples of the link between body states and cognitive & affective states in social interaction, including the impact on motor control, memory, judging facial expressions, reasoning, and general effectiveness in performing tasks.

9.2.1   The Prospective Nature of Helping

For one agent to be able to help another agent, it must first infer the other agent's intentions. We sometimes call this reading intentions. This in itself is a complex problem. It can be addressed in two phases: reading low-level intentions associated with movements (e.g. predicting what someone is reaching for) and reading high-level intentions associated with actions (e.g. predicting why someone is reaching for that object).[7] Instrumental helping requires one agent to understand the goal of another agent by inferring its intention, to recognize that it can't achieve it without assistance, and then to act to provide the necessary help (e.g., picking up something for a person whose hands are full).

The second form of helping — collaboration — is more complicated and focuses on mutual helping where two agents work together to achieve a common goal. It requires the two agents to share their intentions, to agree on the goal, share attention, and engage in joint action. Collaboration requires complex interaction over and above the ability to engage in instrumental helping. It involves the establishment of shared goals and shared intentions[8] and it requires subtle adjustment of actions when the two agents are in physical contact such as when handing items to each other or carrying objects together.

The key to both types of helping is the prospective nature of cognition. Reading intentions, both low-level and high-level, is an exercise in prospection: you have to predict what another agent is going to do but hasn't done yet. Equally, helping someone to do something is prospective because you need to envisage their goal and plan what to do to help the other person achieve it. When collaborating, prospection also allows an agent (robot or human) to establish a compatible representation of some goal that is to be achieved by working together and a shared perspective on what both agents need to do to achieve it together.

9.2.2   Learning to Help and Be Helped

Learning to help is not as easy as it sounds. It takes several years for human infants to develop the requisite abilities, as

[7] Elisabeth Pacherie argues that three different levels of intentions can be distinguished: (1) distal intentions, where the goal-directed action to be performed is specified in cognitive terms with reference to the agent's environment; (2) proximal intentions, where the action is specified in terms of bodily action and the associated perceptual consequence; and (3) motor intentions, where the action is specified in terms of the motor commands and the impact on the agent's sensors. For more details, see her article "The Phenomenology of Action: A Conceptual Framework" [376].

[8] Michael Tomasello and Malinda Carpenter argue that *shared intentionality*, i.e. a collection of social-cognitive and social-motivational skills that allow two or more participants engaged in collaborative activity to share psychological states with one another, plays a crucial role in the development of human infants. In particular, it allows them to transform an ability to follow another agent's gaze into an ability to jointly pay attention to something, to transform social manipulation into cooperative communication, group activity into collaboration, and social learning into instructed learning [377].

we already saw in Chapter 6, Table 6.1, on development and learning.

During the first year of life the progressive acquisition of motor skills determines the development of the ability to understand the intentions of other agents, from anticipating the goal of simple movements to the understanding of more complex goals. At the same time, the ability to infer what another agent is focussing their attention on and the ability to interpret emotional expressions begins to improve substantially.

Around 14 to 18 months of age children begin to exhibit instrumental helping behaviour, i.e. they display spontaneous, unrewarded helping behaviours when another person is unable to achieve his goal.[9] For example, an infant, on seeing another infant unable to reach something it wants to play with, will move it closer to her or him, without the promise of any reward. This is a critical stage in the development of a capacity for collaborative behaviour, a process that progresses past three and four years of age.

Around 2 years of age children start to solve simple cooperation tasks together with adults.[10] This phase of development sees the beginning of shared intentionality where a child and an adult form a shared goal and both engage in joint activity. It also seems that children seem to be motivated not just by the goal but by the cooperation itself, i.e. the social aspect of the interaction.

The ability to cooperate with peers and become a social partner in joint activities develops over the second and third years of life as social understanding increases.[11] More complex collaboration, which necessitates the sharing of intentions and joint coordination of actions, appears at about three years of age when children master more difficult cooperation tasks such as those involving complementary roles for the two partners in a collaborative task.[12]

At three years of age, children begin to develop the ability to cooperate by coordinating two complementary actions. By three-and-a-half years of age children quickly master the task, can deal effectively with the roles in the task being reversed, and can even teach new partners.[13]

The motives which drive instrumental helping are simpler

[9] In "The Roots of Human Altruism" [275], Felix Warneken and Michael Tomasello argue that young children are naturally altruistic and have an innate propensity to help others instrumentally, even when no reward is offered; see Section 9.4 below.

[10] A study by Felix Warneken and colleagues on cooperative activities in young children [378] showed that the ability to engage in non-ritualized cooperative interactions with adults begins to emerge in between 18 and 24 months.

[11] Celia Brownell and colleagues [379] describe significant differences between one- and two-year children in their degree of coordinated activity and level of cooperation in achieving shared goals.

[12] A study of joint action coordination in 2½- and 3-year-old children by Marlene Meyer and colleagues [380] shows the dramatic improvement in the older children in their ability to establish well-coordinated joint action with an adult partner.

[13] A study by Jennifer Ashley and Michael Tomasello [381] shows how children develop the capacity to cooperate by coordinating two complementary actions. 2-year-old children never became independently proficient. 2½- and 3-year-old children showed some proficiency, but mainly as individuals and without much behavioural coordination. However, 3½-year-old children quickly master the set task.

than those of collaborative behaviours: they are based on wanting to see the goal completed or wanting to perceive pleasure in the human at being able to complete it. In this case, the motivational focus is solely on the needs of the second agent. The needs of the first agent don't enter into the equation. The motives underlying collaborative behaviour are more complicated. In this case, the intentions and the goals have to be shared and the motivational focus is on the needs of both agents.

9.3 Reading Intentions and Theory of Mind

As we have said on several occasions, prospection is the very essence of cognitive behaviour: anticipating the need to take action, predicting what action is most appropriate, and prospectively guiding actions during their execution. When actions take on a social aspect, a new dimension of complexity is introduced because other cognitive agents behave in a qualitatively different way to inanimate physical objects. Action becomes interaction.[14] Now, not only does a cognitive agent have to anticipate the behaviour of inanimate objects, it also has to anticipate the behaviour of other cognitive agents. The complexity of this arises because of the very fact that cognitive agents act prospectively and, thus, a cognitive agent must anticipate the actions of an agent that itself is already anticipating what it is going to do. To put it another way, in social cognition, an agent must anticipate the intentions of other agents: it must predict what they will do and possibly why they want to do it.

Infants as young as 18 months understand that there is a difference between the behaviour of inanimate objects and people. They begin to differentiate between physical movements and actions. In particular, they begin to attach a deeper meaning to the movements of people, a meaning which attributes a psychological perspective to their movements, treating them as actions with intended goals, and situating people in a psychological framework involving goals and intentions. Indeed, by 18 months, infants have developed some understanding of intentions.[15]

The ability to infer intentions is closely linked to what is known as a *theory of mind*.[16] We met this concept already in

[14] Tarja Susi and Tom Ziemke provide a neat characterization of interaction in a social context: "Interaction is the reciprocal action where one individual's action may influence and modify the behaviour of another individual" [372].

[15] The ability to infer intentions is related to an agent's capacity to predict the goal of an action being performed by another agent. This capacity is often explained with reference to the mirror neuron system (see Chapter 5, Sections 5.6, and Chapter 6, Section 6.1.2) because it explicitly links the perception of the goal of an action with the agent's own ability to perform that action. This is supported by a study which shows that infants only develop the ability to predict the goals of other people's simple actions after they can perform that action themselves [382].

[16] Read Andrew Meltzoff's article "Understanding the Intentions of Others: Re-Enactment of Intended Acts by 18-Month-Old Children," one of the landmark studies on the development of children's ability to understand the intentions of other people and form a *theory of mind* [273].

Chapter 6, Section 6.1.2, and Chapter 7, Section 7.4, where we defined it to be the capacity by which one agent is able to take a perspective on someone else's situation.[17] It is one of four forms of internal simulation (see Section 7.4). To have a theory of mind means to have the ability to infer what someone else is thinking and wants to do. As we noted in Chapter 6, Section 6.1.2 and Sidenote 15, the innate ability to imitate forms the basis for the development of a person's ability to form a theory of mind. The link between them is the ability of an agent to infer the intentions of another agent. When imitating adults, infants as young as 18 months of age can not only replicate the actions of the adult (and remember: actions are focussed on goals, not just bodily movements) when successfully performing a task but they can also persist in trying to achieve the goal of the action even when the adult is unsuccessful in performing the task. In other words, the infant can read the intention of the adult and infer the unseen goal implied by the unsuccessful attempts.[18]

We saw above that young children differentiate between the behaviour of inanimate and animate objects, attributing mental states to the animate objects. In fact, such is the importance of biological motion to social cognition (as we saw above in Section 9.2) that if an inanimate object, even two-dimensional shapes such as triangles, exhibit movements that are animate or biological — self-propelled, non-linear paths with sudden changes in velocity — humans cannot resist atttributing intentions, emotions, and even personality traits to that inanimate object.[19] In the same way, humans also infer different types of intention depending on whether they are interpreting movements (lower level intentions) or actions (higher level). Whereas movement intention refers to *what* physical state is intended by a certain action, e.g., inferring the end location of a specific observed movement — if the hand moves into the direction of a cup, it is likely that the agent intends to grasp that cup — higher conceptual level intention refers to *why* that specific action is being executed and the motives underlying the action, e.g., the agent might be thirsty and want a drink. This should bring to mind the distinction we made between the concrete movements comprising an action and the higher-order conceptual goals of an action

[17] A *theory of mind* is defined by Andrew Meltzoff as "the understanding of others as psychological beings having mental states such as beliefs, desires, emotions, and intentions" [273].

[18] Andrew Meltzoff and Jean Decety summarize the link between imitation and theory of mind (which they also refer to as *mentalizing*) as follows "Evidently, young toddlers can understand our goals even if we fail to fulfil them. They choose to imitate what we meant to do, rather than what we mistakenly did do." [291], p. 496. As we already noted in Chapter 6, Sidenote 15, they also remark that "In ontogeny, infant imitation is the seed and the adult theory of mind is the fruit."

[19] A classic paper by Fritz Heider and Marianne Simmel in 1944, "An experimental study of apparent behaviour" [383], demonstrates that humans interpret certain types of movements combinations (e.g. successive movement with momentary contact, simultaneous movement with prolonged contact, simultaneous movements without contact, successive movements without contact) as acts of animated beings, mainly people, even when these movements are exhibited by simple two dimensional shapes such as circles and triangles. Furthermore, humans even attribute motives to these acts.

in the previous chapter where we discussed ideo-motor theory (see Chapter 8, Section 8.5.1 and Sidenote 17).

So, how do human infer the intentions of others from their actions? The internal simulation we discussed in Chapter 7, Sections 7.4 and 7.5, is a possible mechanism.[20] The key idea is that the ability to infer the intentions of another agent from observations of their actions might actually be based on the same mechanism that predicts the consequences of the agent's own actions based on its own intentions. As we saw in Chapter 7, Section 7.5, cognitive systems make these predictions by internal simulation using forward models that take either overt or covert motor commands as input and produce as output the likely sensory consequences of carrying out those commands. When a cognitive system observes another agent's actions, the same mechanism can operate provided that the internal simulation mechanism is able to associate observed movements (and not just self-generated motor commands) and likely, i.e. intended, sensory consequences. We saw previously that this is exactly what the ideo-motor principle suggests (Chapter 8, Section 8.5.1) and what the mirror-neuron system provides (in Chapter 5, Section 5.6). In Chapter 6, Section 6.1.2 and Sidenote 15, we discussed the role of imitation in the development of an ability to infer the intentions of others and to empathize with them, providing a mechanism for the developing of a theory of mind. Because ideo-motor models focus on goals and intentions, and because they use a common joint representation that embraces both perception and action, they provide a plausible mechanism for effecting imitation: when an agent sees another agent's actions and the consequences of these actions, the representations of its own actions that would produce the same consequences are activated. By exploiting internal simulation, when an agent just sees another agent's action, not only are the actions activated in it but so too are the consequences of those actions, and hence the intention of the actions can be inferred. With a suitably-sophisticated joint representation and internal simulation mechanism, both low-level movement intentions and high-level action intentions can be accommodated.

While we have discussed intentionality and theory of mind

[20] The review article "From the perception of action to the understanding of intention" by Sarah-Jayne Blakemore and Jean Decety gives a succinct account of the link between biological motion, simulation theory, and the inference of intentions [384]. For an in-depth discussion of a computational approach to intention recognition, see "Towards computational models of intention detection and intention prediction" by Elisheva Bonchek-Dokow and Gal Kaminka [385].

mainly from the perspective of development in children, the ability to infer intentions is very important for artificial cognitive systems, particularly for robots that need to interact naturally with humans, on human terms. As we have just noted, internal simulation is an important mechanism for inferring intentions. In Chapter 7, Sections 7.5.3 and 7.5.4, we discussed the HAMMER architecture and how it uses forward and inverse models to do internal simulation. In fact, HAMMER can also be used to give robots the ability to read intentions in exactly the way suggested in the preceding paragraph: by internal simulation to form a theory of mind and exploiting the multiple pairs of inverse and forward models as a correlate of the mirror neuron system and a realization of ideo-motor theory.[21]

9.4   Instrumental Helping

Young children — as young as one year old — are naturally altruistic and display an innate propensity to help others even when no reward is offered.[22] This altruistic behaviour can take many forms, such as comforting another person who appears to be distressed or helping someone achieve something that they can't do on their own, such as picking up something they've dropped when their hands are full. The latter is referred to as instrumental helping as it focuses on helping another individual achieve their instrumental goal. Thus, the primary goal in instrumental helping is to assist the other individual achieve their goals, even in the absence of any benefit from doing so and sometimes because there is no benefit in doing so. Such behaviour is sometimes referred to as prosocial behaviour[23], in contrast to anti-social behaviour, because they are directed at benefiting another person.

Instrumental helping has two components: a cognitive one and an emotional one. The cognitive component is concerned with recognizing what the other agent's goal is — what they are trying to do — and the reason they can't achieve it on their own. The motivational component is what drives the helping agent to act in the first place. This could be the desire to see the second agent achieve the goal or, alternatively, the desire to see

[21] In addition to reviewing several computational approaches to recognizing actions and intentions, Yiannis Demiris's article "Prediction of intent in robotics and multi-agent systems" [386] also explains how the HAMMER architecture can be used to infer intentions. He notes that predicting and recognizing intentions in situations where there are groups of agents is particularly challenging because the cognitive system has to do more than track and predict the actions of individual agents, it also has to infer the joint intention of the entire group and this may not simply be "the sum of the intentions of the individual agent." It is also necessary to recognize the position of each agent in the social structure of the group. Again, this is a difficult challenge because an agent may play more than one role in the group. The application of the HAMMER architecture to infer the intentions of a disabled wheelchair user is described in his article "Knowing when to assist: developmental issues in lifelong assistive robotics" [387].

[22] This section on *instrumental helping* is a very brief summary of the "The Roots of Human Altruism" by Felix Warneken and Michael Tomasello [275].

[23] The term *prosocial behaviour* was coined by Lauren Wispé in 1972 [388].

the second agent exhibit pleasure at achieving the goal.

The ability to engage in instrumental helping develops with time. 14-month-old infants can help others in situations where the task is relatively simple, e.g. helping with out-of-reach objects, whereas 18-month-old infants engage in instrumental helping in situations where the cognitive task is more complicated.

As already mentioned, rewards are not necessary and the availability of rewards does not increase the incidence of helping. Indeed, rewards can sometimes undermine the motivation to help. Infants are willing to help several times and will even continue to help even if the cost of helping is increased.

Infants as young as 12 months of age offer comfort to other individuals who appear sad or in distress. This is referred to as emotional helping and is motivated by a desire to change the emotional or psychological state of the other agent for the better (as opposed to a desire to see them achieve a goal, as in the case of instrumental helping). It is an open question whether or not young infants combine the motives underlying emotional helping with goal-fulfilment motives during instrumental helping.

9.5   Collaboration

Progressing from instrumental helping to collaboration, the situation becomes more complicated. Here we are dealing with *joint cooperative action*, or *joint action* for short, sometimes referred to as *shared cooperative activity*. Agents that engage in joint action share the same goal, intend to act together, and coordinate their actions to achieve their shared goal through joint attention. That sounds fairly straightforward but as we unwrap each of these issues — joint action, shared intentions, shared goals, and joint attention — the picture becomes complicated because of the interdependencies between them. For example, joint action requires a shared intention, a shared goal, and joint attention when executing the joint action; shared intention includes shared goals; and joint attention is effectively perception that is guided by shared intention and is goal-directed.

To take part in collaborative activities requires an ability to read intentions and infer goals (as was the case in instrumental helping) but it also requires a unique motivation to share psychological states with other agents. By shared intentionality we mean: "collaborative actions in which participants have a shared goal (shared commitment) and coordinated action roles for pursuing that shared goal."[24] What is significant is that the goals and intentions of each agent involved in the collaboration must include something of the goals and intentions of the other agent and something of its own goals and intentions. In other words, the intention is a joint intention and the associated actions are joint actions. This differentiates collaboration from instrumental helping and, as we have said, makes it more complicated. Furthermore, each agent understands both roles of the interaction and so can help the other agent if required. Critically, agents not only choose their own action plan, but also represent the other agent's action plan in its own motor system to enable coordination in the sense of who is doing what and when.

9.5.1 Joint Action

There are at least six degrees of freedom in joint action.[25] These include the number of participants involved, the nature of the relationship between the participants (e.g. peer-to-peer or hierarchical), whether or not the roles are interchangeable, whether the interaction is physical or virtual, whether or not the participants' association is temporary or more long-lasting, and whether or not the interaction is regulated by organizational or cultural norms. In the following, we will assume physical joint action between two peers that temporarily collaborate on a shared goal.

Joint action, or shared cooperative activity, has three essential characteristics:[26]

1. Mutual responsiveness;
2. Commitment to joint activity;
3. Commitment to mutual support.

Let's assume there are two agents engaged in a shared cooperative activity. Each agent must be mutually responsive to the intentions and actions of the other and each must know that the other is trying to be similarly responsive. Consequently, each agent behaves in a way that is guided partially by the behaviour

[24] This quote is take from "Understanding and sharing intentions: The origins of cultural cognition" by Michael Tomasello and colleagues [361].

[25] The six dimensions of joint action are suggested by Elisabeth Pacherie in her article "The Phenomenology of Joint Action: Self-Agency vs. Joint-Agency" [389].

[26] The three characteristics of a *shared cooperative activity* are described in detail in Michael Bratman's article of the same title [390]. Philip Cohen and Hector Levesque address similar issues in their theory of teamwork [391]. They do so in the context of designing artificial agents that can engage in joint action, setting out the conditions that need to be fulfilled for a group of agents to exhibit joint commitment and joint intention. Bratman's account of joint action has been subject to some criticism in that it appears to require sophisticated shared intentionality and an adult-level theory of mind. Yet, as we have seen, young children develop a capability for joint action. An alternative account that doesn't require sophisticated shared intentionality, but only requires shared goals and an understanding of goal-directed actions has been proposed; see "Joint Action and Development" by Stephen Butterfill [392] and "Intentional joint agency: shared intention lite" by Elisabeth Pacherie [393].

of the other agent. This is different from instrumental helping where the helping agent is responsive to the intentions of the agent that needs help but not the other way round.

Each agent must also be committed to the activity in which they are engaged. This means that both agents have the same intention but they need not have the same reason for engaging in the activity. This is a subtle point: it means that the outcome of the collaboration is the same for both agents but the reason for adopting the goal of achieving that outcome need not be the same. If a cognitive robot and a disabled person collaborate to do the laundry, the outcome — the goal — may be a wardrobe full of clean clothes but the reason the person has the goal is to have a fresh shirt to wear in the morning whereas the reason the robot has the goal may just be to keep the house clean and uncluttered.

Finally, each agent must be commited to supporting the efforts of the other to play their role in the joint activity. This characteristic complements the mutual responsiveness by requiring that each agent will in fact provide any help the other agent requires. It says that each agent treats this collaborative mutual support as a priority activity: even if there are other activities that are competing for the attention of each agent, they will still pay attention to the shared cooperative activity they are both engaged in.

Shared intentions are essential for joint action and the intentions of each agent must interlock: each agent must intend that the shared activity be fulfilled in part by the other agent and that their individual activities — both planned actions and actual actions when being executed — mesh together in a mutually-supportive manner.

9.5.2 Shared Intentions

A shared intention — sometimes called *we-intention*, *collective intention*, or *joint intention* — is not simply a collection of individual intentions, even when those individual intentions are supplemented by beliefs or knowledge that both participating agents share. There is more.[27]

An agent with an individual intention represents the overall

[27] Michael Tomasello's and Malinda Carpenter's article "Shared intentionality" [377] provides a good summary of the importance of shared intentionality and its role in the development of socio-cognitive skills such as collaboration and joint attention. An earlier paper by Michael Tomasello and colleagues, "Understanding and sharing intentions: the origins of cultural cognition" [361], discusses the uniqueness of shared intentionality in humans and explains its development on the basis of a peculiarly-human motivation to share emotions, experience, and activities and a more general motivation to understand others as animate, goal-directed, and intentional. An example of how artificial cognitive systems can exploit these ideas can be found in Peter Ford Dominey's and Felix Warneken's paper "The basis of shared intentions in human and robot cognition." Based on findings in computational neuroscience (e.g. the mirror neuron system) and developmental psychology, it describes how representations of shared intentions allow a robot to cooperate with a human [394].

goal and the action plan by which it will achieve that goal and, furthermore, this plan is to be performed by the agent alone. That much is clear. However, agents with a shared intention (and engaged in a joint action) represent the overall shared goal between them but only their own partial sub-plans.[28] Each individual agent with a shared intention does not need to know the other agent's partial plan. However, they do need to share the overall goal. When it comes to the realization of a shared intention and the execution of a joint action, the agent must also factor in the real-time coordination of their individual activities. In this case, each agent must also represent its own actions and their predicted consequences *and* the goals, intentions, actions and predicted consequences of the other agent (just as we discussed above in Section 9.3). Furthermore, each agent must represent the effect that their actions have on the other agent, it must have at least a partial representation of how component actions combine to achieve the overall goal, it must be able to predict the effects of their joint actions so that it can monitor progress towards the overall goal and adjust its actions to help the other agent if necessary (just as we discussed in the previous section).

It is apparent that, in carrying out a joint intention and executing a joint action, both agents must establish a shared perceptual representation. This is where joint attention (in the sense of perception guided by shared intention) comes in.

9.5.3 *Joint Attention*

Social cognition, in general, and collaborative behaviours, in particular, depend on the participating agents to establish *joint attention*.[29]

Joint attention involves much more than two agents looking at the same thing.[30] The essence of joint attention lies in the relationship between intentionality and attention. This provides the basis for a definition of joint attention as "(1) a coordinated and collaborative coupling between intentional agents where (2) the goal of each agent is to attend to the same aspect of the environment."[31] Joint attention, then, requires shared intentionality.

[28] The characterization of *shared intention* in this section is adapted from the treatment by Elisabeth Pacherie [389]. She identifies three levels of shared intentions (shared distal intentions, shared proximal intentions, and coupled motor intentions) which are extensions of her characterization of individual intentions [376] mentioned in Sidenote 7 above. The description of shared intentions in this section omits this distinction but nevertheless begins with characteristics of shared distal intentions before considering the constraints related to the coordination of activity during the real-time execution of joint action, i.e, the constraints that are entailed by shared proximal intentions and coupled motor intentions.

[29] For a comprehensive overview of joint attention, see the survey by Frédéric Kaplan and Verena Hafner [263] which addresses the topic from the perspectives of both developmental psychology and computation modelling. This section closely follows their treatment of the topic.

[30] Michael Tomasello and Malinda Carpenter note that joint attention "is not just two people experiencing the same thing at the same time, but rather it is two people experiencing the same thing at the same time and *knowing together that they are doing this*" [377].

[31] This definition of joint attention is taken from Frédéric Kaplan's and Verena Hafner's article [263].

Furthermore, the participating agents must be engaged in collaborative intentional action. During this collaboration, each agent must monitor, understand, and direct the attentional behaviour of the other agent, and significantly, both agents must be aware that this is going on.

Joint attention is an on-going mutual activity that is carried on throughout the collaborative process to monitor and direct the attention of the other agent. In a sense, joint attention is, itself, a joint activity.

At least four skills need to be recruited by a cognitive agent to achieve joint attention. First, the agent must be able to detect and track the attentional behaviour of the other agent (we are assuming that there are just two agents involved in joint attention here but of course there could be more). Second, the agent must be able to influence the attentional behaviour of the other agent, possibly by using gestures such as pointing or by use of appropriate words. Third, the agent must be able to engage in social coordination to manage the interaction, using techniques such as taking turns or swapping roles, for example. Finally, the agent must be aware that the other agent has intentions (which, as we noted, could be different provided the goal is the same). That is, the agent must be capable of intentional understanding: it must be able to interpret and predict the behaviour of the other agent in terms of the actions required to reach the shared goal.

9.6   Development and Interaction Dynamics

In Chapter 6 on development and learning, Section 6.1.2 and Sidenote 11, we remarked very briefly on two quite different theories of cognitive development, one due to Jean Piaget and the other to Lev Vygotsky.[32] We opened this chapter by remarking that our treatment so far in this primer presented cognition as a somewhat "lonely" process, focussing on an individual agent and that agent's more or less autonomous striving to make sense of its world: acting, anticipating the need for actions, and prospectively choosing and guiding the actions most appropriate for its own goals. We then spent the rest of the chapter balancing this picture, focussing on the social elements of cognition

[32] The explanation of the contrast between the cognitive development theories of Jean Piaget (1896–1980) and Lev Vygotsky (1896–1934) in this section follows the treatment in Kerstin Dautenhahn's and Aude Billard's paper "Studying Robot Social Cognition withing a Developmental Psychology Framework" [285]. There is a summary of Vygotsky's psychology and a discussion of its relevance to artificial cognitive systems, specifically cognitive humanoid robots, in "Social Situatedness of Natural and Artificial Intelligence: Vygotsky and Beyond" by Jessica Lindblom and Tom Ziemke [242]. For a more in-depth overview of Piagetian and Vygotskian developmental psychology, see Chapter 2 of Jessica Lindblom's book *Embodied Social Cognition*, [395]. See also Piaget's and Vygotsky's landmark books *The Construction of Reality in the Child* [396] and *Mind in Society: The Development of Higher Psychological Processes* [397], respectively.

that are required for effective interaction with other agents — inferring intentions, instrumental helping, cooperation, leading to the essence of social cognition: collaborative behaviour and its constituent aspects of shared intentionality, joint action, and joint attention. To a large extent, as we remarked in Section 6.1.2, the agent-centred perspective reflects the cognitive development theory of Jean Piaget while the social cognition perspective mirrors the theory of Lev Vygotsky. So, it is fitting to bring the final chapter of this primer to a close by referring again to these two theories: it serves to remind us of the importance of development in cognition and it highlights the complementary nature of the motivations that drive cognitive development: the individual and the social. It also reprises some of the issues we raised through out the book: the importance of interaction, the role of autonomy, and the primacy of prospection in cognition.

Piaget's theory of developmental psychology focusses on the spontaneous development of the child as he or she interacts with the world. While the child's social context might help promote development, the child's own activity is primary: in exploring and understanding his or her own capabilities and in determining his or her relationship to the world around it. In Piaget's theory, a child goes through several stages of development: a sensorimotor stage up to two years of age, a preoperational stage from two to seven years of age, a concrete operational state from seven to eleven years of age, and finally a formal operational state from eleven years on. Each stage build on — scaffolds — what the child has learned for itself in the previous phases of development. However, the primary building blocks of the child's knowledge are based on his or her own first-hand experiential interaction with the object and people around it. Piaget's theory is effectively a constructivist one (see Section 8.3.4) whereby the child constructs his or her own understanding of the world through first-hand exploration of that world rather than watching other people do things or being told about things by other people.

On the other hand, Vygotsky's position is that the social context is the essential element in development.[33] Teaching plays a pivotal role and a child's cognitive development is heavily in-

[33] Jessica Lindblom points out that Piaget did in fact recognize the importance of the social and cultural dimension to development but notes that it affects more the speed of development rather than the direction of development [395].

fluenced by the cultural norms in which the child is immersed. These cultural and social patterns determine the way the child will develop and the way the child will learn to understand the world around it. In the same way, social dynamics are an important aspect of this development. The coordination of movements between agents as they interact plays a key role in the development of cognitive skills, both because of the natural need to synchronize activities during cooperative activities and because of the need to coordinate the agent's goals and intentions with those of the other agent with which it is interacting. Among many other things, Vygotsky introduced the concept of the *zone of proximal development* to characterize the situation where children develop skills that are just beyond their current capabilites, doing so with the help of another agent or in collaboration with another agent. In a neat twist, the zone of proximal development has also been used to identify the degree of assistance that an artificial cognitive system should provide to a disabled user of a wheelchair in order to balance the current needs of the user with the longer-term rehabilitation potential provided by adaptive assistive technology.[34]

Perhaps it is best to view the Piagetian and Vygotskian positions as complementary rather than incompatible alternatives. After all, we saw shades of both positions throughout this book. For example, in Chapter 6, we emphasized that there are two types of motivation that drive development in a child: the exploratory and the social. In Chapter 2 we highlighted that a striving to make sense of the world is a key characteristic of the emergent paradigm of cognitive science and, especially, the enactive stance on cognition while in Chapters 2 and 5 we noted that the meaning of knowledge emerges through interaction: it is negotiated by two or more agents as they interact and what something means is agreed by consensus.

Ultimately, cognitive development is a journey of discovery: to determine what matters and what does not matter to an agent. Discovering what matters allows the agent to act prospectively, both to help itself and to help others, and in so doing to construct an understanding of the world, an understanding that manages an effective tradeoff between being autonomous and

[34] The incorporation of Vygotsky's concept of the *zone of proximal development* to modulate the level of assistance provided by a cognitive system to a disabled user is described in Yiannis Demiris's article "Knowing when to assist: developmental issues in lifelong assistive robotics" [387]. This cognitive system uses the HAMMER architecture discussed in Chapter 7, Section 7.5.3, to infer the intentions of a disabled user of a wheelchair.

surrendering some of that autonomy to protect the social environment upon which the agent depends. Piaget's position reflects the child's part in the process of discovery through spontaneous exploration; Vygotsky's position recognizes the essential role of the social interaction in guiding that journey and determining what each act of discovery reveals. In the end, this process of discovery, this development of cognition, gives the agent — biological or artificial — the ability to anticipate the need for action and the flexibility to adapt as the world throws unexpected events in its path.

References

[1] http://www.commsp.ee.ic.ac.uk/~mcpetrou/iron.html.

[2] http://en.wikipedia.org/wiki/Maria_Petrou.

[3] J. Maitin-Shepard, M. Cusumano-Towner, J. Lei, and P. Abbeel. Cloth grasp point detection based on multiple-view geometric cues with application to robotic towel folding. In *International Conference on Robotics and Automation ICRA*, pages 2308–2315, 2010.

[4] Willow Garage. The PR2 robot. http://www.willowgarage.com/pages/pr2/overview, 2013.

[5] A. Morse and T. Ziemke. On the role(s) of modelling in cognitive science. *Pragmatics & Cognition*, 16(1):37–56, 2008.

[6] T. C. Scott-Phillips, T. E. Dickins, and S. A. West. Evolutionary theory and the ultimate-proximate distinction in the human behavioural sciences. *Perspectives on Psychological Science*, 6(1):38–47, 2011.

[7] N. Tinbergen. On the aims and methods of ethology. *Zeitschrift für Tierpsychologie*, 20:410–433, 1963.

[8] E. Mayr. *Animal species and evolution*. Harvard University Press, Cambridge, MA, 1963.

[9] D. Vernon. Cognitive system. In K. Ikeuchi, editor, *Computer Vision: A Reference Guide*, pages 100–106. Springer, 2014.

[10] P. Medawar. *Pluto's Republic*. Oxford University Press, 1984.

[11] http://www.cs.bham.ac.uk/research/projects/cogaff/misc/ meta-requirements.html.

[12] D. Vernon, C. von Hofsten, and L. Fadiga. *A Roadmap for Cognitive Development in Humanoid Robots*, volume 11 of *Cognitive Systems Monographs (COSMOS)*. Springer, Berlin, 2010.

[13] M. H. Bickhard. Autonomy, function, and representation. *Artificial Intelligence, Special Issue on Communication and Cognition*, 17(3-4):111–131, 2000.

[14] H. Maturana and F. Varela. *The Tree of Knowledge — The Biological Roots of Human Understanding*. New Science Library, Boston & London, 1987.

[15] R. J. Brachman. Systems that know what they're doing. *IEEE Intelligent Systems*, 17(6):67–71, December 2002.

[16] A. Sloman. Varieties of affect and the cogaff architecture schema. In *Proceedings of the AISB '01 Symposium on Emotion, Cognition, and Affective Computing*, York, UK, 2001.

[17] R. Sun. The importance of cognitive architectures: an analysis based on clarion. *Journal of Experimental & Theoretical Artificial Intelligence*, 19(2):159–193, 2007.

[18] D. Marr. *Vision*. Freeman, San Francisco, 1982.

[19] T. Poggio. The *levels of understanding* framework, revised. *Perception*, 41:1017–1023, 2012.

[20] D. Marr and T. Poggio. From understanding computation to understanding neural circuitry. In E. Poppel, R. Held, and J. E. Dowling, editors, *Neuronal Mechanisms in Visual Perception*, volume 15 of *Neurosciences Research Program Bulletin*, pages 470–488. 1977.

[21] J. A. S. Kelso. *Dynamic Patterns – The Self-Organization of Brain and Behaviour*. MIT Press, Cambridge, MA, 3rd edition, 1995.

[22] R. Pfeifer and J. Bongard. *How the body shapes the way we think: a new view of intelligence*. MIT Press, Cambridge, MA, 2007.

[23] A. Clark. *Being There: Putting Brain, Body, and World Together Again*. MIT Press, Cambridge, MA, 1997.

[24] A. Clark. Time and mind. *Journal of Philosophy*, XCV(7):354–376, 1998.

[25] N. Wiener. *Cybernetics: or the Control and Communication in the Animal and the Machine*. John Wiley and Sons, New York, 1948.

[26] W. Ross Ashby. *An Introduction to Cybernetics*. Chapman and Hall, London, 1957.

[27] W. S. McCulloch and W. Pitts. A logical calculus of ideas immanent in nervous activity. *Bulletin of Mathematical Biophysics*, 5:115–133, 1943.

[28] J. A. Anderson and E. Rosenfeld, editors. *Neurocomputing: Foundations of Research*. MIT Press, Cambridge, MA, 1988.

[29] W. R. Ashby. *Design for a Brain*. John Wiley & Sons, New York, first edition. edition, 1952.

[30] W. R. Ashby. *Design for a Brain*. John Wiley & Sons, New York, first edition. reprinted with corrections. edition, 1954.

[31] W. R. Ashby. *Design for a Brain*. John Wiley & Sons, New York, second edition. edition, 1960.

[32] F. J. Varela. Whence perceptual meaning? A cartography of current ideas. In F. J. Varela and J.-P. Dupuy, editors, *Understanding Origins – Contemporary Views on the Origin of Life, Mind and Society*, Boston Studies in the Philosophy of Science, pages 235–263, Dordrecht, 1992. Kluwer Academic Publishers.

[33] A. Clark. *Mindware – An Introduction to the Philosophy of Cognitive Science*. Oxford University Press, New York, 2001.

[34] M. Waibel, M. Beetz, J. Civera, R. D'Andrea, J. Elfring, D. Gáalvez-Loópez, K. Häussermann, R. Janssen, J. M. M. Montiel, A. Perzylo, B. Schießlele, M. Tenorth, O. Zweigle, and R. van de Molengraft. Roboearth: A world-wide web for robots. *IEEE Robotics and Automation Magazine*, pages 69–82, June 2011.

[35] A. Newell and H. A. Simon. Computer science as empirical inquiry: Symbols and search. *Communications of the Association for Computing Machinery*, 19:113–126, March 1976. Tenth Turing award lecture, ACM, 1975.

[36] W. J. Freeman and R. Núñez. Restoring to cognition the forgotten primacy of action, intention and emotion. *Journal of Consciousness Studies*, 6(11-12):ix–xix, 1999.

[37] http://www.aaai.org/Conferences/AAAI/2012/ aaai12cognitivecall.php.

[38] http://www.agi-society.org.

[39] http://www.agiri.org/wiki/Artificial_General_Intelligence.

[40] S. Harnad. The symbol grounding problem. *Physica D*, 42:335–346, 1990.

[41] A. Newell. The knowledge level. *Artificial Intelligence*, 18(1):87–127, March 1982.

[42] J. E. Laird, A. Newell, and P. S. Rosenbloom. Soar: an architecture for general intelligence. *Artificial Intelligence*, 33(1–64), 1987.

[43] A. Newell. *Unified Theories of Cognition*. Harvard University Press, Cambridge MA, 1990.

[44] J. Anderson. Cognitive architectures in rational analysis. In K. van Lehn, editor, *Architectures for Intelligence*, pages 1–24. Lawrence Erlbaum Associates, Hillsdale, NJ, 1999.

[45] http://ai.eecs.umich.edu/cogarcho.

[46] E. von Glaserfeld. *Radical Constructivism*. Routeledge-Falmer, London, 1995.

[47] W. D. Christensen and C. A. Hooker. An interactivist-constructivist approach to intelligence: self-directed anticipative learning. *Philosophical Psychology*, 13(1):5–45, 2000.

[48] J. Stewart, O. Gapenne, and E. A. Di Paolo. *Enaction: Toward a New Paradigm for Cognitive Science*. MIT Press, 2010.

[49] J. P. Crutchfield. Dynamical embodiment of computation in cognitive processes. *Behavioural and Brain Sciences*, 21(5):635–637, 1998.

[50] D. A. Medler. A brief history of connectionism. *Neural Computing Surveys*, 1:61–101, 1998.

[51] J. A. Anderson and E. Rosenfeld, editors. *Neurocomputing 2: Directions for Research*. MIT Press, Cambridge, MA, 1991.

[52] P. Smolensky. Computational, dynamical, and statistical perspectives on the processing and learning problems in neural network theory. In P. Smolensky, M. C. Mozer, and D. E. Rumelhart, editors, *Mathematical perspectives on neural networks*, pages 1–15. Erlbaum, Mahwah, NJ, 1996.

[53] P. Smolensky. Computational perspectives on neural networks. In P. Smolensky, M. C. Mozer, and D. E. Rumelhart, editors, *Mathematical perspectives on neural networks*, pages 1–15. Erlbaum, 1996.

[54] P. Smolensky. Dynamical perspectives on neural networks. In P. Smolensky, M. C. Mozer, and D. E. Rumelhart, editors, *Mathematical perspectives on neural networks*, pages 245–270. Erlbaum, 1996.

[55] P. Smolensky. Statistical perspectives on neural networks. In P. Smolensky, M. C. Mozer, and D. E. Rumelhart, editors, *Mathematical perspectives on neural networks*, pages 453–496. Erlbaum, 1996.

[56] M. A. Arbib, editor. *The Handbook of Brain Theory and Neural Networks*. MIT Press, Cambridge, MA, 1995.

[57] RJ. A. Feldman and D. H. Ballard. Connectionist models and their properties. *Cognitive Science*, 6:205–254, 1982.

[58] E. L. Thorndike. *The Fundamentals of Learning*. Teachers College, Columbia University, New York, 1932.

[59] E. L. Thorndike. *Selected Writings from a Connectionist Psychology*. Greenwood Press, New York, 1949.

[60] W. James. *The Principles of Psychology*, volume 1. Harvard University Press, Cambridge, MA, 1890.

[61] D. O. Hebb. *The Organization of Behaviour*. John Wiley & Sons, New York, 1949.

[62] F. Rosenblatt. The perceptron: a probabilistic model for information storage and organization in the brain. *Psychological Review*, 65:386–408, 1958.

[63] O. G. Selfridge. Pandemonium: A paradigm for learning. In D. V. Blake and A. M. Uttley, editors, *Proceedings of the Symposium on Mechanization of Thought Processes*, pages 511–529, London, 1959. H. M. Stationery Office.

[64] B. Widrow and M. E. Hoff. Adaptive switching circuits. In *1960 IRE WESCON Convention Record*, pages 96–104, New York, 1960.

[65] M. Minsky and S. Papert. *Perceptrons: An Introduction to Computational Geometry*. MIT Press, Cambridge, MA, 1969.

[66] J. Pollack. No harm intended: Marvin L. Minsky and Seymour A. Papert. Perceptrons: An introduction to computational geometry, expanded edition. *Journal of Mathematical Psychology*, 33(3):358–365, 1989.

[67] G. E. Hinton and J. A. Anderson, editors. *Parallel models of associative memory*. Lawrence Eralbaum Associates, Hillsdale, N.J.:, 1981.

[68] G. A. Carpenter and S. Grossberg. Adaptive resonance theory (ART). In M. A. Arbib, editor, *The Handbook of Brain Theory and Neural Networks*, pages 79–82. MIT Press, Cambridge, MA, 1995.

[69] T. Kohonen. Self-organized formation of topologically correct feature maps. *Biological Cybernetics*, 43:59–69, 1982.

[70] D. E. Rumelhart, J. L. McClelland, and The PDP Research Group, editors. *Parallel Distributed Processing: Explorations in the Microstructure of Cognition*. The MIT Press, Cambridge, 1986.

[71] D. E. Rumelhart, G. E. Hinton, and R. J. Williams. Learning internal representations by error propagation. In D. E. Rumelhart, J. L. McClelland, and The PDP Research Group, editors, *Parallel Distributed Processing: Explorations in the Microstructure of Cognition*, pages 318–362. The MIT Press, Cambridge, 1986.

[72] D. E. Rumelhart, G. E. Hinton, and R. J. Williams. Learning representations by back-propagating erros. *Nature*, 323:533–536, 1986.

[73] P. Werbos. *Beyond regression: new tools for prediction and analysis in the behaviourl sciences*. Masters Thesis. Harvard University, Boston, MA, 1974.

[74] J. J. Hopfield. Neural neural network and physical systems with emergent collective computational abilities. *Proceedings of National Academy of Sciences*, 79(8):2554 – 2588, 1982.

[75] J. Elman. Finding structure in time. *Cognitive Science*, 14:179–211, 1990.

[76] M. I. Jordan. Attractor dynamics and parallelism in a connectionist sequential machine. In *Proceedings of the Eighth Conference of the Cognitive Science Society*, pages 531–546, 1986.

[77] G. E. Hinton and T. J. Sejnowski. Learning and relearning in boltzmann machines. In D. E. Rumelhart, J. L. McClelland, and The PDP Research Group, editors, *Parallel Distributed Processing: Explorations in the Microstructure of Cognition*, pages 282–317, Cambridge, 1986. The MIT Press.

[78] J. Moody and C. J. Darken. Fast learning in networks of locally tuned processing units. *Neural Computation*, 1:281–294, 1989.

[79] J. L. McClelland and T. T. Rogers. The parallel distributed processing approach to semantic cognition. *Nature*, 4:310–322, 2003.

[80] P. Smolensky and G. Legendre. *The Harmonic Mind: From Neural Computation To Optimality-Theoretic Grammar*. MIT Press, 2006.

[81] P. Smolensky. structure and explanation in an integrated connectionist/symbolic cognitive architecture. In C. Macdonald and G. Macdonald, editors, *Connectionism: Debates on psychological explanation*, volume 2, pages 221–290. Basil Blackwell, 1995.

[82] T. van Gelder and R. F. Port. It's about time: An overview of the dynamical approach to cognition. In R. F. Port and T. van Gelder, editors, *Mind as Motion – Explorations in the Dynamics of Cognition*, pages 1–43, Cambridge, Massachusetts, 1995. Bradford Books, MIT Press.

[83] L. Shapiro. *Embodied Cognition*. Routledge, 2011.

[84] E. Thelen and L. B. Smith. *A Dynamic Systems Approach to the Development of Cognition and Action*. MIT Press / Bradford Books Series in Cognitive Psychology. MIT Press, Cambridge, Massachusetts, 1994.

[85] S. Camazine. Self-organizing systems. In *Encyclopedia of Cognitive Science*. Wiley, 2006.

[86] A. Kravchenko. Essential properties of language, or, why language is not a code. *Language Sciences*, 5(29):650–671, 2007.

[87] T. Winograd and F. Flores. *Understanding Computers and Cognition – A New Foundation for Design*. Addison-Wesley Publishing Company, Inc., Reading, Massachusetts, 1986.

[88] J. P. Spencer, M. S. C. Thomas, and J. L. McClelland. *Toward a New Grand Theory of Development? Connectionism and Dynamic Systems Theory Re-Considered*. Oxford University Press, New York, 2009.

[89] G. Schöner. Development as change of dynamic systems: Stability, instability, and emergence. In J. P. Spencer, M. S. C. Thomas, and J. L. McClelland, editors, *Toward a Unified Theory of Development: Connectionism and Dynamic Systems Theory Re-Considered*, New York, 2009. Oxford University Press.

[90] P. Smolensky, M. C. Mozer, and D. E. Rumelhart, editors. *Mathematical perspectives on neural networks*. Erlbaum, 1996.

[91] J. J. Gibson. The theory of affordances. In R. Shaw and J. Bransford, editors, *Perceiving, acting and knowing: toward an ecological psychology*, pages 67–82. Lawrence Erlbaum, 1977.

[92] W. Köhler. *Dynamics in Psychology*. Liveright, New York, 1940.

[93] D. Vernon. Enaction as a conceptual framework for development in cognitive robotics. *Paladyn Journal of Behavioral Robotics*, 1(2):89–98, 2010.

[94] H. Maturana. Biology of cognition. Research Report BCL 9.0, University of Illinois, Urbana, Illinois, 1970.

[95] H. Maturana. The organization of the living: a theory of the living organization. *Int. Journal of Man-Machine Studies*, 7(3):313–332, 1975.

[96] H. R. Maturana and F. J. Varela. *Autopoiesis and Cognition — The Realization of the Living*. Boston Studies on the Philosophy of Science. D. Reidel Publishing Company, Dordrecht, Holland, 1980.

[97] F. Varela. *Principles of Biological Autonomy*. Elsevier North Holland, New York, 1979.

[98] F. Varela, E. Thompson, and E. Rosch. *The Embodied Mind.* MIT Press, Cambridge, MA, 1991.

[99] R. Grush. The emulation theory of representation: motor control, imagery, and perception. *Behavioral and Brain Sciences*, 27:377–442, 2004.

[100] G. Hesslow. Conscious thought as simulation of behaviour and perception. *Trends in Cognitive Sciences*, 6(6):242–247, 2002.

[101] M. P. Shanahan. A cognitive architecture that combines internal simulation with a global workspace. *Consciousness and Cognition*, 15:433–449, 2006.

[102] H. H. Clark. Managing problems in speaking. *Speech Communication*, 15:243–250, 1994.

[103] D. Vernon, G. Metta, and G. Sandini. A survey of artificial cognitive systems: Implications for the autonomous development of mental capabilities in computational agents. *IEEE Transactions on Evolutionary Computation*, 11(2):151–180, 2007.

[104] T. Froese and T. Ziemke. Enactive artificial intelligence: Investigating the systemic organization of life and mind. *Artificial Intelligence*, 173:466–500, 2009.

[105] D. Vernon and D. Furlong. Philosophical foundations of enactive AI. In M. Lungarella, F. Iida, J. C. Bongard, and R. Pfeifer, editors, *50 Years of AI*, volume LNAI 4850, pages 53–62. Springer, Heidelberg, 2007.

[106] W. D. Christensen and C. A. Hooker. Representation and the meaning of life. In H. Clapin, P. Staines, and P. Slezak, editors, *Representation in Mind: New Approaches to Mental Representation*, pages 41–70. Elsevier, Oxford, 2004.

[107] M. P. Shanahan and B. Baars. Applying global workspace theory to the frame problem. *Cognition*, 98(2):157–176, 2005.

[108] G. Granlund. Organization of architectures for cognitive vision systems. In H. I. Christensen and H.-H. Nagel, editors, *Cognitive Vision Systems: Sampling the Spectrum of Approaches*, volume 3948 of *LNCS*, pages 37–56, Heidelberg, 2006. Springer-Verlag.

[109] J. R. Anderson, D. Bothell, M. D. Byrne, S. Douglass, C. Lebiere, and Y. Qin. An integrated theory of the mind. *Psychological Review*, 111(4):1036–1060, 2004.

[110] P. Rosenbloom, J. Laird, and A. Newell, editors. *The Soar Papers: Research on Integrated Intelligence.* MIT Press, Cambridge, Massachusetts, 1993.

[111] J. F. Lehman, J. E. Laird, and P. S. Rosenbloom. A gentle introduction to soar, an architecture for human cognition. In S. Sternberg and D. Scarborough, editors, *Invitation to Cognitive Science, Volume 4: Methods, Models, and Conceptual Issues.* MIT Press, Cambridge, MA, 1998.

[112] J. E. Laird. Extending the soar cognitive architecture. In *Proceedings of the First Conference on Artificial General Intelligence*, pages 224–235, Amsterdam, The Netherlands, 2008. IOS Press.

[113] J. E. Laird. Towards cognitive robotics. In G. R. Gerhart, D. W. Gage, and C. M. Shoemaker, editors, *Proceedings of the SPIE — Unmanned Systems Technology XI*, volume 7332, pages 73320Z–73320Z–11, 2009.

[114] J. E. Laird. *The Soar Cognitive Architecture.* MIT Press, Cambridge, MA, 2012.

[115] J. R. Anderson. Act: A simple theory of complex cognition. *American Psychologist*, 51:355–365, 1996.

[116] R. Sun. A tutorial on CLARION 5.0. In *Cognitive Science Department*. Rensselaer Polytechnic Institute, 2003. http://www.cogsci.rpi.edu/~rsun/sun.tutorial.pdf.

[117] R. Sun. Desiderata for cognitive architectures. *Philosophical Psychology*, 17(3):341–373, 2004.

[118] W. D. Gray, R. M. Young, and S. S. Kirschenbaum. Introduction to this special issue on cognitive architectures and
human-computer interaction. *Human-Computer Interaction*,
12:301–309, 1997.

[119] P. Langley. An adaptive architecture for physical agents. In
*IEEE/WIC/ACM International Conference on Intelligent Agent
Technology*, pages 18–25, Compiegne, France, 2005. IEEE
Computer Society Press.

[120] P. Langley, J. E. Laird, and S. Rogers. Cognitive architectures: Research issues and challenges. *Cognitive Systems
Research*, 10(2):141–160, 2009.

[121] F. E. Ritter and R. M. Young. Introduction to this special issue on using cognitive models to improve interface design.
International Journal of Human-Computer Studies, 55:1–14,
2001.

[122] H. von Foerster. *Understanding Understanding: Essays on
Cybernetics and Cognition*. Springer, New York, 2003.

[123] J. L. Krichmar and G. M. Edelman. Brain-based devices
for the study of nervous systems and the development of
intelligent machines. *Artificial Life*, 11:63–77, 2005.

[124] J. L. Krichmar and G. N. Reeke. The Darwin brain-based
automata: Synthetic neural models and real-world devices.
In G. N. Reeke, R. R. Poznanski, K. A. Lindsay, J. R. Rosen-
berg, and O. Sporns, editors, *Modelling in the neurosciences:
from biological systems to neuromimetic robotics*, pages 613–
638, Boca Raton, 2005. Taylor and Francis.

[125] J. L. Krichmar and G. M. Edelman. Principles underlying
the construction of brain-based devices. In T. Kovacs
and J. A. R. Marshall, editors, *Proceedings of AISB '06 -
Adaptation in Artificial and Biological Systems*, volume 2 of
*Symposium on Grand Challenge 5: Architecture of Brain and
Mind*, pages 37–42, Bristol, 2006. University of Bristol.

[126] J. Weng. Developmental robotics: Theory and experiments. *International Journal of Humanoid Robotics*, 1(2):199–236, 2004.

[127] K. E. Merrick. A comparative study of value systems for self-motivated exploration and learning by robots. *IEEE Transactions on Autonomous Mental Development*, 2(2):119–131, June 2010.

[128] P.-Y. Oudeyer, F. Kaplan, and V. Hafner. Intrinsic motivation systems for autonomous mental development. *IEEE Transactions on Evolutionary Computation*, 11(2):265–286, 2007.

[129] N. Hawes, J. Wyatt, and A. Sloman. An architecture schema for embodied cognitive systems. In *Technical Report CSR-06-12*. University of Birmingham, School of Computer Science, 2006.

[130] N. Hawes and J. Wyatt. Developing intelligent robots with CAST. In *Proc. IROS Workshop on Current Software Frameworks in Cognitive Robotics Integrating Different Computational Paradigms*, 2008.

[131] S. C. Shapiro and J. P. Bona. The GLAIR cognitive architecture. In A. Samsonovich, editor, *Biologically Inspired Cognitive Architectures-II: Papers from the AAAI Fall Symposium, Technical Report FS-09-01*, pages 141–152. AAAI Press, Menlo Park, CA, 2009.

[132] P. Langley. Cognitive architectures and general intelligent systems. *AI Magazine*, 27(2):33–44, 2006.

[133] P. Langley, D. Choi, and S. Rogers. Acquisition of hierarchical reactive skills in a unified cognitive architecture. *Cognitive Systems Research*, 10(4):316–332, 2009.

[134] A. Morse, R. Lowe, and T. Ziemke. Towards an enactive cognitive architecture. In *Proceedings of the First International Conference on Cognitive Systems*, Karlsruhe, Germany, 2008.

[135] T. Ziemke and R. Lowe. On the role of emotion in embodied cognitive architectures: From organisms to robots. *Cognition and Computation*, 1:104–117, 2009.

[136] G. Metta, L. Natale, F. Nori, G. Sandini, D. Vernon, L. Fadiga, C. von Hofsten, J. Santos-Victor, A. Bernardino, and L. Montesano. The iCub Humanoid Robot: An Open-Systems Platform for Research in Cognitive Development. *Neural Networks, special issue on Social Cognition: From Babies to Robots*, 23:1125–1134, 2010.

[137] G. Sandini, G. Metta, and D. Vernon. The icub cognitive humanoid robot: An open-system research platform for enactive cognition. In M. Lungarella, F. Iida, J. C. Bongard, and R. Pfeifer, editors, *50 Years of AI*, volume LNAI 4850, pages 359ñ–370. Springer, Heidelberg, 2007.

[138] J. Weng. A theory of developmental architecture. In *Proceedings of the 3rd International Conference on Development and Learning (ICDL 2004)*, La Jolla, October 2004.

[139] C. Burghart, R. Mikut, R. Stiefelhagen, T. Asfour, H. Holzapfel, P. Steinhaus, and R. Dillman. A cognitive architecture for a humanoid robot: A first approach. In *IEEE-RAS International Conference on Humanoid Robots (Humanoids 2005)*, pages 357–362, 2005.

[140] S. Franklin. A foundational architecture for artificial general intelligence. In B. Goertzel and P. Wang, editors, *Proceeding of the 2007 conference on Advances in Artificial General Intelligence: Concepts, Architectures and Algorithms*, pages 36–54, Amsterdam, 2007. IOS Press.

[141] D. Friedlander and S. Franklin. LIDA and a theory of mind. In *Proceeding of the 2008 conference on Advances in Artificial General Intelligence*, pages 137–148, Amsterdam, 2008. IOS Press.

[142] D. Kraft, E. Başeski, M. Popović, A. M. Batog, A. Kjær-Nielsen, N. Krüger, R. Petrick, C. Geib, N. Pugeault,

M. Steedman, T. Asfour, R. Dillmann, S. Kalkan, F. Wörgötter, B. Hommel, R. Detry, and J. Piater. Exploration and planning in a three-level cognitive architecture. In *Proceedings of the First International Conference on Cognitive Systems*, Karlsruhe, Germany, 2008.

[143] http://bicasociety.org/cogarch/architectures.htm.

[144] W. Duch, R. J. Oentaryo, and M. Pasquier. Cognitive architectures: Where do we go from here? In *Proc. Conf. Artificial General Intelligence*, pages 122–136, 2008.

[145] J. L. Krichmar, D. A. Nitz, J. A. Gally, and G. M. Edelman. Characterizing functional hippocampal pathways in a brain-based device as it solves a spatial memory task. *Proceedings of the National Academy of Science, USA*, 102:2111–2116, 2005.

[146] J. L. Krichmar, A. K. Seth, D. A. Nitz, J. G. Fleisher, and G. M. Edelman. Spatial navigation and causal analysis in a brain-based device modelling cortical-hippocampal interactions. *Neuroinformatics*, 3:197–221, 2005.

[147] A.K. Seth, J.L. McKinstry, G.M. Edelman, and J. L. Krichmar. Active sensing of visual and tactile stimuli by brain-based devices. *International Journal of Robotics and Automation*, 19(4):222–238, 2004.

[148] E. L. Bienenstock, L. N. Cooper, and P. W. Munro. Theory for the development of neuron selectivity: orientation specificity and binocular interaction in visual cortex. *Journal of Neurscience*, 2(1):32–48, 1982.

[149] K. Kawamura, S. M. Gordon, P. Ratanaswasd, E. Erdemir, and J. F. Hall. Implementation of cognitive control for a humanoid robot. *International Journal of Humanoid Robotics*, 5(4):547–586, 2008.

[150] M. A. Boden. Autonomy: What is it? *BioSystems*, 91:305–308, 2008.

[151] T. Froese, N. Virgo, and E. Izquierdo. Autonomy: a review and a reappraisal. In F. Almeida e Costa et al., editor, *Proceedings of the 9th European Conference on Artificial Life: Advances in Artificial Life*, volume 4648, pages 455–465. Springer, 2007.

[152] T. Ziemke. On the role of emotion in biological and robotic autonomy. *BioSystems*, 91:401–408, 2008.

[153] A. Seth. Measuring autonomy and emergence via Granger causality. *Artificial Life*, 16(2):179–196, 2010.

[154] N. Bertschinger, E. Olbrich, N. Ay, and J. Jost. Autonomy: An information theoretic perspective. *Biosystems*, 91(2):331–345, 2008.

[155] W. F. G. Haselager. Robotics, philosophy and the problems of autonomy. *Pragmatics & Cognition*, 13:515–532, 2005.

[156] T. B. Sheridan and W. L. Verplank. Human and computer control for undersea teleoperators. Technical report, MIT Man-Machine Systems Laboratory, 1978.

[157] M. A. Goodrich and A. C. Schultz. Human–robot interaction: A survey. *Foundations and Trends in Human–Computer Interaction*, 1(3):203–275, 2007.

[158] J. M. Bradshaw, P. J. Feltovich, H. Jung, S. Kulkarni, W. Taysom, and A. Uszok. Dimensions of adjustable autonomy and mixed-initiative interaction. In M. Nickles, M. Rovatos, and G. Weiss, editors, *Agents and Computational Autonomy: Potential, Risks, and Solutions*, volume 2969 of *LNAI*, pages 17–39. Springer, Berlin/Heidelberg, 2004.

[159] C. Castelfranchi. Guarantees for autonomy in cognitive agent architecture. In M. J. Woolridge and N. R. Jennings, editors, *Intelligent Agents*. Springer, Berlin/Heidelberg, 1995.

[160] C. Castelfranchi and R. Falcone. Founding autonomy: The dialectics between (social) environment and agent's architecture and powers. In M. Nickles, M. Rovatos, and

G. Weiss, editors, *Agents and Computational Autonomy: Potential, Risks, and Solutions*, volume 2969 of *LNAI*, pages 40–54. Springer, Berlin/Heidelberg, 2004.

[161] C. Carabelea, O. Boissier, and A. Florea. Autonomy in multi-agent systems: A classification attempt. In M. Nickles, M. Rovatos, and G. Weiss, editors, *Agents and Computational Autonomy: Potential, Risks, and Solutions*, volume 2969 of *LNAI*, pages 103–113. Springer, Berlin/Heidelberg, 2004.

[162] A. Meystel. From the white paper explaining the goals of the workshop: "Measuring performance and intelligence of systems with autonomy: Metrics for intelligence of constructed systems". In E. Messina and A. Meystel, editors, *Proceedings of the 2000 PerMIS Workshop*, volume Special Publication 970, Gaithersburg, MD, U.S.A., August 14-16 2000. NIST.

[163] B. Pitzer, M. Styer, C. Bersch, C. DuHadway, and J. Becker. Towards perceptual shared autonomy for robotic mobile manipulation. In *IEEE International Conference on Robotics and Automation (ICRA)*, pages 6245–6251, 2011.

[164] J. W. Crandall, M. A. Goodrich, D. R. Olsen, and C. W. Nielsen. Validating human-robot interaction schemes in multi-tasking environments. *IEEE Transactions on Systems, Man, and Cybernetics: Part A — Systems and Humans*, 35(4):438–449, 2005.

[165] W. D. Christensen and C. A. Hooker. Representation and the meaning of life. In *Representation in Mind: New Approaches to Mental Representation*. The University of Sydney, June 2000.

[166] W. B. Cannon. Organization of physiological homeostasis. *Physiological Reviews*, 9:399–431, 1929.

[167] C. Bernard. Les phénomènes de la vie. Paris, 1878.

[168] A. R. Damasio. *Looking for Spinoza: Joy, sorrow and the feeling brain*. Harcourt, Orlando, Florida, 2003.

[169] P. Sterling. Principles of allostasis. In J. Schulkin, editor, *Allostasis, Homeostasis, and the Costs of Adaptation*. Cambridge University Press., Cambridge, England, 2004.

[170] P. Sterling. Allostasis: A model of predictive behaviour. *Physiology and Behaviour*, 106(1):5–15, 2012.

[171] I. Muntean and C. D. Wright. Autonomous agency, AI, and allostasis — a biomimetic perspective. *Pragmatics & Cognition*, 15(3):485–513, 2007.

[172] M. H. Bickhard and D. T. Campbell. Emergence. In P. B. Andersen, C. Emmeche, N. O. Finnemann, and P. V. Christiansen, editors, *Downward Causation*, pages 322–348. University of Aarhus Press, Aarhus, Denmark, 2000.

[173] J. O. Kephart and D. M. Chess. The vision of autonomic computing. *IEEE Computer*, 36(1):41–50, 2003.

[174] P. Horn. Autonomic computing: IBM's perspective on the state of information technology, October 2001.

[175] IBM. An architectureal blueprint for autonomic computing. White paper, 2005.

[176] J. L. Crowley, D. Hall, and R. Emonet. Autonomic computer vision systems. In *The 5th International Conference on Computer Vision Systems*, 2007.

[177] D. Vernon. Reconciling autonomy with utility: A roadmap and architecture for cognitive development. In A. V. Samsonovich and K. R. Jóhannsdóttir, editors, *Proc. Int. Conf. on Biologically-Inspired Cognitive Architectures*, pages 412–418. IOS Press, 2011.

[178] E. Di Paolo and H. Iizuka. How (not) to model autonomous behaviour. *BioSystems*, 91:409–423, 2008.

[179] C. E. Shannon. A mathematical theory of communication. *Bell System Technical Journal*, 27:379–423, 623–656, 1948.

[180] C. Granger. Investigating causal relations by econometric models and cross-spectral methods. *Econometrica*, 37:424–438, 1969.

[181] A. Seth. Granger causality. *Scholarpedia*, 2:1667, 2007.

[182] M. Schillo and K. Fischer. A taxonomy of autonomy in multiagent organisation. In M. Nickles, M. Rovatos, and G. Weiss, editors, *Agents and Computational Autonomy: Potential, Risks, and Solutions*, volume 2969 of *LNAI*, pages 68–82. Springer, Berlin/Heidelberg, 2004.

[183] X. Barandiaran. Behavioral adaptive autonomy. A milestone in the Alife route to AI? In *Proceedings of the 9th International Conference on Artificial Life*, pages 514–521, Cambridge: MA, 2004. MIT Press.

[184] M. Schillo. Self-organization and adjustable autonomy: Two sides of the same medal? *Connection Science*, 14:345—359, 2003.

[185] H. Hexmoor, C. Castelfranchi, and R. Falcone, editors. *Agent Autonomy*. Kluwer, Dordrecht, The Netherlands, 2003.

[186] K. Ruiz-Mirazo and A. Moreno. Basic autonomy as a fundamental step in the synthesis of life. *Artificial Life*, 10(3):235–259, 2004.

[187] K. S. Barber and J. Park. Agent belief autonomy in open multi-agent systems. In M. Nickles, M. Rovatos, and G. Weiss, editors, *Agents and Computational Autonomy: Potential, Risks, and Solutions*, volume 2969 of *LNAI*, pages 7–16. Springer, Berlin/Heidelberg, 2004.

[188] E. A. Di Paolo. Unbinding biological autonomy: Francisco Varela's contributions to artificial life. *Artificial Life*, 10(3):231–233, 2004.

[189] C. Melhuish, I. Ieropoulos, J. Greenman, and I. Horsfield. Energetically autonomous robots: Food for thought. *Autonomous Robotics*, 21:187–198, 2006.

[190] I. Ieropoulos, J. Greenman, C. Melhuish, and I. Horsfield. Microbial fuel cells for robotics: Energy autonomy through artificial symbiosis. *ChemSusChem*, 5(6):1020–1026, 2012.

[191] A. Moreno, A. Etxeberria, and J. Umerez. The autonomy of biological individuals and artificial models. *BioSystems*, 91:309—319, 2008.

[192] B. Sellner, F. W. Heger, L. M. Hiatt, R. Simmons, and S. Singh. Coordinated multi-agent teams and sliding autonomy for large-scale assembly. *Proceedings of the IEEE*, 94(7), 2006.

[193] R. Chrisley and T. Ziemke. Embodiment. In *Encyclopedia of Cognitive Science*, pages 1102–1108. Macmillan, 2002.

[194] M. L. Anderson. Embodied cognition: A field guide. *Artificial Intelligence*, 149(1):91–130, 2003.

[195] G. Piccinini. The mind as neural software? Understanding functionalism, computationalism, and computational functionalism. *Philosophy and Phenomenological Research*, 81(2):269ñ–311, September 2010.

[196] A. Clark. The dynamical challenge. *Cognitive Science*, 21:461–481, 1997.

[197] E. Thelen. Time-scale dynamics and the development of embodied cognition. In R. F. Port and T. van Gelder, editors, *Mind as Motion – Explorations in the Dynamics of Cognition*, pages 69–100, Cambridge, Massachusetts, 1995. Bradford Books, MIT Press.

[198] A. Riegler. When is a cognitive system embodied? *Cognitive Systems Research*, 3(3):339–348, 2002.

[199] E. Thelen and L. B. Smith. *A Dynamic Systems Approach to the Development of Cognition and Action*. MIT Press, Cambridge, Massachusetts, 3rd edition, 1998.

[200] E. Thelen and L. B. Smith. Development as a dynamic system. *Trends in Cognitive Sciences*, 7:343–348, 2003.

[201] R. A. Wilson and L. Foglia. Embodied cognition. In E. N. Zalta, editor, *The Stanford Encyclopedia of Philosophy*. 2011.

[202] R. Brooks. Elephants don't play chess. *Robotics and Autonomous Systems*, 6:3–15, 1990.

[203] A. D. Wilson and S. Golonka. Embodied cognition is not what you think it is. *Frontiers in Psychology*, 4, 2013.

[204] T. McGreer. Passive dynamic walking. *International Journal of Robotics Research*, 9:62–82, 1990.

[205] M. Wilson. Six views of embodied cognition. *Psychonomic Bulletin & Review*, 9(4):625–636, 2002.

[206] A. Clark and J. Toribio. Doing without representing? *Synthese*, 101:401–431, 1994.

[207] L. W. Barsalou, P. M. Niedenthal, A. Barbey, and J. Ruppert. Social embodiment. In B. Ross, editor, *The Psychology of Learning and Motivation*, volume 43, pages 43–92. Academic Press, San Diego, 2003.

[208] L. Craighero, M. Nascimben, and L. Fadiga. Eye position affects orienting of visuospatial attention. *Current Biology*, 14:331–333, 2004.

[209] L. Craighero, L. Fadiga, G. Rizzolatti, and C. A. Umiltà. Movement for perception: a motor-visual attentional effect. *Journal of Experimental Psychology: Human Perception and Performance*, 1999.

[210] J. R. Lackner. Some proprioceptive influences on the perceptual representation of body shape and orientation. *Brain*, 111:281–297, 1988.

[211] G. Rizzolatti and L. Fadiga. Grasping objects and grasping action meanings: the dual role of monkey rostroventral premotor cortex (area F5). In G. R. Bock and J. A. Goode, editors, *Sensory Guidance of Movement, Novartis Foundation Symposium 218*, pages 81–103. John Wiley and Sons, Chichester, 1998.

[212] V. Gallese, L. Fadiga, L. Fogassi, and G. Rizzolatti. Action recognition in the premotor cortex. *Brain*, 119:593–609, 1996.

[213] G. Rizzolatti, L. Fadiga, V. Gallese, and L. Fogassi. Premotor cortex and the recognition of motor actions. *Cognitive Brain Research*, 3:131–141, 1996.

[214] G. Rizzolatti and L. Craighero. The mirror neuron system. *Annual Review of Physiology*, 27:169–192, 2004.

[215] C. von Hofsten. An action perspective on motor development. *Trends in Cognitive Sciences*, 8:266–272, 2004.

[216] S. Thill, D. Caligiore, A. M. Borghi, T. Ziemke, and G. Baldassarre. Theories and computational models of affordance and mirror systems: An integrative review. *Neuroscience and Biobehavioral Reviews*, 37:491–521, 2013.

[217] A. D. Barsingerhorn, F. T. J. M. Zaal, J. Smith, and G.-J. Pepping. On the possibilities for action: The past, present and future of affordance research. *AVANT*, III, 2012.

[218] H. Svensson, J. Lindblom, and T. Ziemke. Making sense of embodied cognition: Simulation theories of shared neural mechanisms for sensorimotor and cognitive processes. In T. Ziemke, J. Zlatev, and R. M. Frank, editors, *Body, Language and Mind*, volume 1: Embodiment, pages 241–269. Mouton de Gruyter, Berlin, 2007.

[219] M. Bickhard. Is embodiment necessary? In P. Calvo and T. Gomila, editors, *Handbook of Cognitive Science: An Embodied Approach*, pages 29–40. Elsevier, 2008.

[220] T. Ziemke. Disentangling notions of embodiment. In R. Pfeifer, M. Lungarella, and G. Westermann, editors, *Workshop on Developmental and Embodied Cognition*, Edinburgh, UK., July 2001.

[221] T. Ziemke. Are robots embodied? In C. Balkenius, J. Zlatev, K. Dautenhahn, H. Kozima, and C. Breazeal, editors, *Proceedings of the First International Workshop on Epigenetic Robotics — Modeling Cognitive Development in Robotic Systems*, volume 85 of *Lund University Cognitive Studies*, pages 75–83, Lund, Sweden, 2001.

[222] T. Ziemke. What's that thing called embodiment? In R. Alterman and D. Kirsh, editors, *Proceedings of the 25th Annual Conference of the Cognitive Science Society*, Lund University Cognitive Studies, pages 1134–1139, Mahwah, NJ, 2003. Lawrence Erlbaum.

[223] K. Dautenhahn, B. Ogden, and T. Quick. From embodied to socially embedded agents – implications for interaction-aware robots. *Cognitive Systems Research*, 3(3):397–428, 2002.

[224] N. Sharkey and T. Ziemke. Life, mind and robots — the ins and outs of embodied cognition. In S. Wermter and R. Sun, editors, *Hybrid Neural Systems*, pages 314–333. Springer Verlag, 2000.

[225] N. Sharkey and T. Ziemke. Mechanistic vs. phenomenal embodiment: Can robot embodiment lead to strong AI? *Cognitive Systems Research*, 3(3):251–262, 2002.

[226] J. von Uexküll. *Umwelt und Innenwelt der Tiere.* Springer, Berlin, 1921.

[227] J. von Uexküll. A stroll through the worlds of animals and men —- a picture book of invisible worlds. In C. H. Schiller, editor, *Instinctive behavior — The development of a modern concept*, pages 5–80. International Universities Press, New York, 1957.

[228] B. Ogden, K. Dautenhahn, and P. Stribling. Interactional structure applied to the identification and generation of visual interactive behaviour: Robots that (usually) follow the rules. In I. Wachsmuth and T. Sowa, editors, *Gesture and Sign Languages in Human-Computer Interaction*, volume LNAI 2298 of *Lecture Notes LNAI*, pages 254–268. Springer, 2002.

[229] T. Ziemke. Embodied AI as science: Models of embodied cognition, embodied models of cognition, or both? In F. Iida, R. Pfeifer, L. Steels, and Y. Kuniyoshi, editors, *Embodied Artificial Intelligence*, volume LNAI 3139, pages 27–36. Springer-Verlag, 2004.

[230] R. Núñez. Could the future taste purple? Reclaiming mind, body and cognition. *Journal of Consciousness Studies*, 6(11–12):41–60, 1999.

[231] A. Clark. An embodied cognitive science? *Trends in Cognitive Sciences*, 9:345–351, 1999.

[232] Y. Demiris and B. Khadhouri. Hierarchical attentive multiple models for execution and recognition (HAMMER). *Robotics and Autonomous Systems*, 54:361–369, 2006.

[233] Y. Demiris, L. Aziz-Zahdeh, and J. Bonaiuto. Information processing in the mirror neuron system in primates and machines. *Neuroinformatics*, 12(1):63–91, 2014.

[234] M. Anderson. How to study the mind: An introduction to embodied cognition. In F. Santoianni and C. Sabatano, editors, *Brain Development in Learning Environments: Embodied and Perceptual Advancements*, pages 65–82. Cambridge Scholars Press, 2007.

[235] X. Huang and J. Weng. Novelty and reinforcement learning in the value system of developmental robots. In *Proc. Second International Workshop on Epigenetic Robotics: Modeling Cognitive Development in Robotic Systems (EPIROB '02)*, pages 47–55, Edinburgh, Scotland, 2002.

[236] X. Huang and J. Weng. Inherent value systems for autonomous mental development. *International Journal of Humanoid Robotics*, 4:407—433, 2007.

[237] D. Parisi. Internal robotics. *Connection Science*, 16(4):325–338, 2004.

[238] M. Stapleton. Steps to a "Properly Embodied" cognitive science. *Cognitive Systems Research*, 22–23:1–11, 2013.

[239] W. J. Clancey. *Situated Cognition: On Human Knowledge and Computer Representations*. Cambridge University Press, Cambridge MA, 1997.

[240] P. Robbins and M. Aydede, editors. *Cambridge Handbook of Situated Cognition*. Cambridge University Press, Cambridge, UK, 2008.

[241] A. Clark. *Microcognition: Philosophy, Cognitive Science and Parallel Distributed Processing*. MIT Press, 1989.

[242] J. Lindblom and T. Ziemke. Social situatedness of natural and artificial intelligence: Vygotsky and beyond. *Adaptive Behavior*, 11(2):79–96, 2003.

[243] L. W. Barsalou. Grounded cognition. *Annu. Rev. Psychol.*, 59(11):II.1–II.29, 2008.

[244] L. W. Barsalou. Grounded cognition: Past, present, and future. *Topics in Cognitive Science*, 2:716–724, 2010.

[245] A. Clark and D. Chalmers. The extended mind. *Analysis*, 58:10–23, 1998.

[246] A. Clark. *Supersizing the Mind: Embodiment, Action, and Cognitive Extension*. Oxford University Press, 2008.

[247] A. Clark. Précis of supersizing the mind: Embodiment, action, and cognitive extension (oxford university press, ny, 2008). *Philosophical Studies*, 152:413—416, 2011.

[248] J. Fodor. Where is my mind? *London Review of Books*, 31(3):13–15, 2009.

[249] E. Hutchins. *Cognition in the Wild*. MIT Press, Cambridge, MA, 1995.

[250] J. Hollan, E. Hutchins, and D. Kirsh. Distributed cognition: Toward a new foundation for human-computer interaction research. *ACM Transactions on Computer-Human Interaction*, 7(2):174–196, 2000.

[251] E. Hutchins. How a cockpit remembers its speed. *Cognitive Science*, 19:265–288, 1995.

[252] P. Calvo and T. Gomila, editors. *Handbook of Cognitive Science: An Embodied Approach*. Elsevier, 2008.

[253] L. Shapiro. The embodied cognition research programme. *Philosophy Compass*, 2(2):338—346, 2007.

[254] T. Ziemke. Introduction to the special issue on situated and embodied cognition. *Cognitive Systems Research*, 3(3):271–274, 2002.

[255] http://www.eucognition.org/index.php?page=tutorial-on-embodiment.

[256] T. Farroni, G. Csibra, F. Simion, and M. H. Johnson. Eye contact detection in humans from birth. *Proceeding of the National Academy of Sciences (PNAS)*, 99:9602–9605, 2002.

[257] T. Farroni, M. H. Johnson, E. Menon, L. Zulian, D. Faraguna, and G. Csibra. Newborns' preference for face-relevant stimuli: Effect of contrast polarity. *Proceeding of the National Academy of Sciences (PNAS)*, 102:17245—17250, 2005.

[258] F. Simion, L. Regolin, and H. Bulf. A predisposition for biological motion in the newborn baby. *Proceeding of the National Academy of Sciences (PNAS)*, 105(2):809–813, 2008.

[259] M. H. Johnson, S. Dziurawiec, H. D. Ellis, and J. Morton. Newborns' preferential tracking of face-like stimuli and its subsequent decline. *Cognition*, 40:1–19, 1991.

[260] E. Valenza, F. Simion, V. M. Cassia, and C. Umiltà. Face preference at birth. *J. Exp. Psychol. Hum. Percept. Perform.*, 22:892–903, 1996.

[261] J. Alegria and E. Noirot. Neonate orientation behaviour towards human voice. *Int. J. Behav. Dev.*, 1:291–312, 1978.

[262] J. M. Haviland and M. Lelwica. The induced affect response: 10-week-old infants' responses to three emotion expressions. *Developmental Psychology*, 23:97–104, 1987.

[263] F. Kapland and V. Hafner. The challenges of joint attention. *Interaction Studies*, 7(2):135–169, 2006.

[264] G. Young-Browne, H. M. Rosenfeld, and F. D. Horowitz. Infant discrimination of facial expressions. *Child Development*, 48(2):555–562, 1977.

[265] D. P. F. Montague and A. S. Walker-Andrews. Peekaboo: A new look at infants' perception of emotion expression. *Developmental Psychology*, 37:826—838, 2001.

[266] R. Flom and L. E. Bahrick. The development of infant discrimination of affect in multimodal and unimodal stimulation: The role of intersensory redundancy. *Developmental Psychology*, 43:238–252, 2007.

[267] G. Butterworth. The ontogeny and phylogeny of joint visual attention. In A. Whiten, editor, *Natural theories of mind: Evolution, development, and simulation of everyday mind*. Blackwell, 1991.

[268] G. Butterworth and N. Jarrett. What minds have in common is space: Spatial mechanisms serving joint visual attention in infancy. *British Journal of Developmental Psychology*, 9:55–72, 1991.

[269] R. Brooks and A. N. Meltzoff. The importance of eyes: How infants interpret adult looking behavior. *Developmental Psychology*, 38(6):958–966, 2002.

[270] A. L. Woodward and J. J. Guajardo. Infants' understanding of the point gesture as an object-directed action. *Cognitive Development*, 17:1061–1084, 2002.

[271] Gredebäck and A. Melinder. Infants' understanding of everyday social interactions: a dual process account. *Cognition*, 114(2):197–206, 2010.

[272] B. M. Repacholi and A. Gopnik. Early reasoning about desires: Evidence from 14- and 18-month-olds. *Developmental Psychology*, 33:12–21, 1997.

[273] A. N. Meltzoff. Understanding the intentions of others: Re-enactment of intended acts by 18-month-old children. *Developmental Psychology*, 31:838—850, 1995.

[274] F. Bellagamba and M. Tomasello. Re-enacting intended acts: Comparing 12- and 18-month-olds. *Infant Behavior and Development*, 22:277–282, 1999.

[275] F. Warneken and M. Tomasello. The roots of human altruism. *British Journal of Psychology*, 100(3):455–471, 2009.

[276] M. Lungarella, G. Metta, R. Pfeifer, and G. Sandini. Developmental robotics: A survey. *Connection Science*, 15:151–190, 2003.

[277] M. Asada, K. Hosoda, Y. Kuniyoshi, H. Ishiguro, T. Inui, Y. Yoshikawa, M. Ogino, and C. Yoshido. Cognitive developmental robotics: A survey. *IEEE Transactions on Autonomous Mental Development*, 1(1):12–34, May 2009.

[278] G. Sandini, G. Metta, and J. Konczak. Human sensori-Motor development and artificial systems. In *Proceedings of AIR & IHAS*, Japan, 1997.

[279] E. S. Reed. *Encountering the world: towards an ecological psychology*. Oxford University Press, New York, 1996.

[280] G. Lintern. Encountering the world: Toward an ecological psychology by Edward S. Reed. *Complexity*, 3(6):61–63, 1998.

[281] C. von Hofsten. Action, the foundation for cognitive development. *Scandinavian Journal of Psychology*, 50:617–623, 2009.

[282] C. von Hofsten. *Action Science: The emergence of a new discipline*, chapter Action in infancy: a foundation for cognitive development, pages 255–279. MIT Press, 2013.

[283] J. P. Forgas, K. D. Kipling, and S. M. Laham, editors. *Social Motivation*. Cambridge University Press, 2005.

[284] C. von Hofsten. On the development of perception and action. In J. Valsiner and K. J. Connolly, editors, *Handbook of Developmental Psychology*, pages 114–140. Sage, London, 2003.

[285] K. Dautenhahn and A. Billard. Studying robot social cognition within a developmental psychology framework. In *Proceedings of Eurobot 99: Third European Workshop on Advanced Mobile Robots*, pages 187–194, Switzerland, 1999.

[286] A. Billard. Imitation. In M. A. Arbib, editor, *The Handbook of Brain Theory and Neural Networks*, pages 566–569. MIT Press, Cambridge, MA, 2002.

[287] A. N. Meltzoff. The elements of a developmental theory of imitation. In A. N. Meltzoff and W. Prinz, editors, *The Imitative Mind: Development, Evolution, and Brain Bases*, pages 19–41. Cambridge University Press, Cambridge, 2002.

[288] A. N. Meltzoff and M. K. Moore. Imitation of facial and manual gestures by human neonates. *Science*, 198:75–78, 1977.

[289] A. N. Meltzoff and M. K. Moore. Explaining facial imitation: A theoretical model. *Early Development and Parenting*, 6:179–192, 1997.

[290] R. Rao, A. Shon, and A. Meltzoff. A Bayesian model of imitation in infants and robots. In K. Dautenhahn and C. Nehaniv, editors, *Imitation and Social Learning in Robots, Humans, and Animals: Behaviour, Social and Communicative Dimensions*. Cambridge University Press, 2004.

[291] A. N. Meltzoff and J. Decety. What imitation tells us about social cognition: a rapprochement between developmental psychology and cognitive neuroscience. *Philosophical Transactions of the Royal Society of London: Series B*, 358:491–500, 2003.

[292] http://en.wikipedia.org/wiki/George_E._P._Box.

[293] http://ai.eecs.umich.edu/cogarch2/prop/monotonicity.html.

[294] T. Mitchell. *Machine Learning*. McGraw Hill, 1997.

[295] K. Doya. What are the computations of the cerebellum, the basal ganglia and the cerebral cortex? *Neural Networks*, 12:961–974, 1999.

[296] K. Doya. Complementary roles of basal ganglia and cerebellum in learning and motor control. *Current Opinion in Neurobiology*, 10:732–739, 2000.

[297] M. Harmon and S. Harmon. http://www.dtic.mil/cgi-bin/GetTRDoc?AD=ADA323194, 1997.

[298] Z. Ghahramani. Unsupervised learning. volume LNAI 3176 of *Advanced Lectures on Machine Learning*. Springer-Verlag, 2004.

[299] J. L. McClelland, B. L. NcNaughton, and R. C. O'Reilly. Why there are complementary learning systems in the hippocampus and neocortex: insights from the successes and failures of connectionist models of learning and memory. *Psychological Review*, 102(3):419–457, 1995.

[300] B. D. Argall, S. Chernova, M. Veloso, and B. Browning. A survey of robot learning from demonstration. *Robotics and Autonomous Systems*, 57:469–483, 2009.

[301] A. Sloman and J. Chappell. The altricial-precocial spectrum for robots. In *IJCAI '05 – 19th International Joint Conference on Artificial Intelligence*, Edinburgh, 30 July – 5 August 2005.

[302] R. S. Johansson, G. Westling, A. Bäckström, and J. R. Flanagan. Eye-hand coordination in object manipulation. *Journal of Neuroscience*, 21(17):6917–6932, 2001.

[303] J. R. Flanagan and R. S. Johansson. Action plans used in action observation. *Nature*, 424(769–771), 2003.

[304] E. S. Spelke. Core knowledge. *American Psychologist*, pages 1233–1243, November 2000.

[305] L. Feigenson, S. Dehaene, and E. S. Spelke. Core systems of number. *Trends in Cognitive Sciences*, 8:307–314, 2004.

[306] R. Fox and C. McDaniel. The perception of biological motion by human infants. *Science*, 218:486–487, 1982.

[307] L. Hermer and E. S. Spelke. Modularity and development: the case of spatial reorientation. *Cognition*, 61(195–232), 1996.

[308] R. F. Wang and E. S. Spelke. Human spatial representation: insights from animals. *Trends in Cognitive Sciences*, 6:376–382, 2002.

[309] R. Wood, P. Baxter, and T. Belpaeme. A review of long-term memory in natural and synthetic systems. *Adaptive Behavior*, 20(2):81–103, 2012.

[310] L. Squire. Memory systems of the brain: a brief history and current perspective. *Neurobiology of Learning and Memory*, 82:171–177, 2004.

[311] J. Fuster. Network memory. *Trends in Neurosciences*, 20(10):451–459, 1997.

[312] P. Baxter and W. Browne. Memory as the substrate of cognition: a developmental cognitive robotics perspective. In *Proc. Epigenetic Robotics 10*, Örenäs Slott, Sweden, November 2010.

[313] N. Cowan. Evolving conceptions of memory storage, selective attention, and their mutual constraints within the human information-processing system. *Psychological Bulletin*, 104:163–191, 1988.

[314] D. Durstewitz, J. K. Seamans, and T. J. Sejnowski. Neurocomputational models of working memory. *Nature Neuroscience Supplement*, 3:1184–1191, 2000.

[315] G. Ryle. *The concept of mind*. Hutchinson's University Library, London, 1949.

[316] E. Tulving. Episodic and semantic memory. In E. Tulving and W. Donaldson, editors, *Organization of memory*, pages 381–403. Academic Press, New York, 1972.

[317] E. Tulving. Précis of *elements of episodic memory. Behavioral and Brain Sciences*, 7:223–268, 1984.

[318] M. E. P. Seligman, P. Railton, R. F. Baumeister, and C. Sripada. Navigating into the future or driven by the past. *Perspectives on Psychological Science*, 8(2):119–141, 2013.

[319] E. Tulving. *Elements of Episodic Memory*. Oxford University Press, 1983.

[320] A. Berthoz. *The Brain's Sense of Movement*. Harvard University Press, Cambridge, MA, 2000.

[321] D. L. Schacter and D. R. Addis. Constructive memory — the ghosts of past and future: a memory that works by piecing together bits of the past may be better suited to simulating future events than one that is a store of perfect records. *Nature*, 445:27, 2007.

[322] L. Carroll. *Through the Looking-Glass*. 1872.

[323] K. Downing. Predictive models in the brain. *Connection Science*, 21:39–74, 2009.

[324] D. L. Schacter, D. R. Addis, and R. L. Buckner. Episodic simulation of future events: Concepts, data, and applications. *Annals of the New York Academy of Sciences*, 1124:39–60, 2008.

[325] R. L. Buckner and D. C. Carroll. Self-projection and the brain. *Trends in Cognitive Sciences*, 11:49–57, 2007.

[326] Y. Østby, K. B. Walhovd, C. K. Tamnes, H. Grydeland, L. G. Westlye, and A. M. Fjell. Mental time traval and default-mode network functional connectivity in the developing brain. *PNAS*, 109(42):16800–16804, 2012.

[327] C. M. Atance and D. K. O'Neill. Episodic future thinking. *Trends in Cognitive Sciences*, 5(12):533–539, 2001.

[328] K. K. Szpunar. Episodic future throught: An emerging concept. *Perspectives on Psychological Science*, 5(2):142–162, 2010.

[329] D. L. Schacter and D. R. Addis. The cognitive neuroscience of constructive memory: Remembering the past and imagining the future. *Philosophical Transactions of the Royal Society B*, 362:773–786, 2007.

[330] C. M. Atance and D. K. O'Neill. The emergence of episodic future thinking in humans. *Learning and Motivation*, 36:126–144, 2005.

[331] D. T. Gilbert and T. D. Wilson. Prospection: Experiencing the future. *Science*, 317:1351–1354, 2007.

[332] R. Lowe and T. Ziemke. The feeling of action tendencies: on the emotional regulation of goal-directed behaviours. *Frontiers in Psychology*, 2:1–24, 2011.

[333] R. Picard. Affective computing. Technical Report 321, MIT Media Lab, Cambridge, MA, 1995.

[334] R. Picard. *Affective Computing*. MIT Press, Cambridge, MA, 1997.

[335] K. R. Scherer, T. Bänziger, and E. Roesch. *A Blueprint for Affective Computing*. Oxford University Press, Oxford, UK, 2010.

[336] G. Hesslow. The current status of the simulation theory of cognition. *Brain Research*, 1428:71–79, 2012.

[337] H. Svensson, S. Thill, and T. Ziemke. Dreaming of electric sheep? Exploring the functions of dream-like mechanisms in the development of mental imagery simulations. *Adaptive Behavior*, 21:222–238, 2013.

[338] H. Svensson, A. F. Morse, and T. Ziemke. Representation as internal simulation: A minimalistic robotic model. In N. Taatgen and H. van Rijn, editors, *Proceedings of the Thirty-first Annual Conference of the Cognitive Science Society*, pages 2890–2895, Austin, TX, 2009. Cognitive Science Society.

[339] S. T. Moulton and S. M. Kosslyn. Imagining predictions: Mental imagery as mental emulation. *Philosophical Transactions of the Royal Society B*, 364:1273–1280, 2009.

[340] S. Wintermute. Imagery in cognitive architecture: Representation and control at multiple levels of abstraction. *Cognitive Systems Research*, 19–20:1–29, 2012.

[341] M. Iacoboni. Imitation, empathy, and mirror neurons. *Annual Review of Psychology*, 60:653–670, 2009.

[342] B. Hommel, J. Müsseler, G. Aschersleben, and W. Prinz. The theory of event coding (TEC): A framework for perception and action planning. *Behavioral and Brain Sciences*, 24:849–937, 2001.

[343] H. Gravato Marques and O. Holland. Architectures for functional imagination. *Neurocomputing*, 72(4-6):743–759, 2009.

[344] S. Thill and H. Svensson. The inception of simulation: a hypothesis for the role of dreams in young children. In L. Carlson, C. Hoelscher, and T. F. Shipley, editors, *Proceedings of the Thirty-Third Annual Conference of the Cognitive Science Society*, pages 231–236, Austin, TX, 2011. Cognitive Science Society.

[345] D. Wolpert, R. C. Miall, and M. Kawato. Internal models in the cerebellum. *Trends in Cognitive Sciences*, 2(9):338–347, 1998.

[346] D. Wolpert, J. Diedrichsen, and J. R. Flanagan. Principles of sensorimotor learning. *Nature Reviews Neuroscience*, 12:39–751, December 2011.

[347] M. Kawato. Internal models for motor control and trajectory planning. *Current Opinion in Neurobiology*, 9(6):718–727, 1999.

[348] A. Nuxoll, D. Tecuci, W. C. Ho, and N. Wang. Comparing forgetting algorithms for artificial episodic memory systems. In M. Lim and W. C. Ho, editors, *Proceedings of the*

remembering who we are — human memory for artificial agents symposium AISB, pages 14–20, 2010.

[349] J. T. Wixted. The psychology and neuroscience of forgetting. *Annu. Ref. Psychol.*, 55:235–269, 2004.

[350] P. beim Graben, T. Liebscher, and J. Kurths. Neural and cognitive modeling with networks of leaky integrator units. In P. beim Graben, C. Zhou, M. Thiel, and J. Kurths, editors, *Lectures in Supercomputational Neurosciences*, pages 195–223. Springer, Berlin, 2008.

[351] J. Staddon. The dynamics of memory in animal learning. In M. Sabourin, F. Craik, and M. Robert, editors, *Advances in psychological science*, volume 2: Biological and cognitive aspects, pages 259–274. Taylor and Francis, Hove, England, 1988.

[352] H. Svensson and T. Ziemke. Embodied representation: What are the issues? In B. G. Bara, L. Barsalou, and M. Bucciarelli, editors, *Proceedings of the Twenty-Seventh Annual Conference of the Cognitive Science Society*, pages 2116–2121, Mahwah, NJ, 2005. Erlbaum.

[353] P. Haselager, A. de Groot, and H. van Rappard. Representationalism vs. anti-representationalism: a debate for the sake of appearance. *Philosophical Psychology*, 16:5–23, 2005.

[354] A. Riegler. The constructivist challenge. *Constructivist Foundations*, 1(1):1–8, 2005.

[355] E. von Glaserfeld. Aspectos del constructivismo radical (Aspects of radical constructivism). In M. Pakman, editor, *Construcciones de la experiencia humana*, pages 23–49. Barcelona, Spain, 1996.

[356] G. Lakoff and R. Núñez. *Where Mathematics Comes from: How the Embodied Mind Brings Mathematics into Being*. Basic Books, 2000.

[357] Silvia Coradeschi and Alessandro Saffiotti. An introduction to the anchoring problem. *Robotics and Autonomous Systems*, 43:85–96, 2003.

[358] A. Sloman.
http://www.cs.bham.ac.uk/research/projects/cogaff/
misc/talks/models.pdf.

[359] A. Stock and C. Stock. A short history of ideo-motor
action. *Psychological research*, 68(2–3):176–188, 2004.

[360] M. E. Bratman. Intention and personal policies. In J. E.
Tomberlin, editor, *Philosophical Perspectives*, volume 3.
Blackwell.

[361] M. Tomasello, M. Carpenter, J. Call, T. Behne, and H. Moll.
Understanding and sharing intentions: the origins of cultural cognition. *Behavioral and Brain Sciences*, 28(5):675–735,
2005.

[362] S. Ondobaka and H. Bekkering. Hierarchy of idea-guided
action and perception-guided movement. *Frontiers in
Cognition*, 3:1–5, 2012.

[363] N. Krüger, C. Geib, J. Piater, R. Petrickb, M. Steedman,
F. Wörgötter, A. Ude, T. Asfour, D. Kraft, D. Omrčen,
A. Agostini, and Rüdiger Dillmann. Object–action com-
plexes: Grounded abstractions of sensory–motor processes.
Robotics and Autonomous Systems, 59:740–757, 2011.

[364] M. Tenorth, A. Clifford Perzylo, R. Lafrenz, and M. Beetz.
Representation and exchange of knowledge about actions,
objects, and environments in the roboearth framework.
*IEEE Transactions on Automation Science and Engineering
(T-ASE)*, 10(3):643–651, July 2013.

[365] M. Tenorth, U. Klank, D. Pangercic, and M. Beetz. Web-
enabled robots. *IEEE Robotics and Automation Magazine*,
pages 58–68, June 2011.

[366] M. Tenorth and M. Beetz. KnowRob — Knowledge processing for autonomous personal robots. In *Proc. IEEE/RSJ
International Conference on Intelligent Robots and Systems*,
pages 4261–4266, 2009.

[367] J. Kuffner.   Cloud enabled robots. http://www.scribd.com/doc/47486324/Cloud-Enabled-Robots.

[368] K. Goldberg and B. Kehoe. Cloud robotics and automation: A survey of related work. Technical Report Technical Report No. UCB/EECS-2013-5, University of California at Berkeley, 2013.

[369] A. Billard, S. Calinon, R. Dillmann, and S. Schaal. Robot programming by demonstration. In *Springer Handbook of Robotics*, pages 1371–1394. 2008.

[370] R. Dillmann, T. Asfour, M. Do, R. Jäkel, A. Kasper, P. Azad, A. Ude, S. Schmidt-Rohr, and M. Lösch. Advances in robot programming by demonstration. *Künstliche Intelligenz*, 24(4):295–303, 2010.

[371] M. Riley, A. Ude, C. Atkeson, and G. Cheng. Coaching: An approach to efficiently and intuitively create humanoid robot behaviors. In *IEEE-RAS Conference on Humanoid Robotics*, pages 567–574, 2006.

[372] T. Susi and T. Ziemke. Social cognition, artefacts, and stigmergy: A comparative analysis of theoretical frameworks for the understanding of artefact-mediated collaborative activity. *Journal of Cognitive Systems Research*, 2:273–290, 2001.

[373] J. K. Tsotsos. *A Computational Perspective on Visual Attention*. MIT Press, 2011.

[374] U. Frith and S-J. Blakemore. Social cognition. In M. Kenward, editor, *Foresight Cognitive Systems Project*. Foresight Directorate, Office of Science and Technology, 1 Victoria Street, London, SW1H 0ET, United Kingdom, 2006.

[375] M. A. Pavlova. Biological motion processing as a hallmark of social cognition. *Cerebral Cortex*, 22(981–995), 2012.

[376] E. Pacherie. The phenomenology of action: A conceptual framework. *Cognition*, 107:179–217, 2008.

[377] M. Tomasello and M. Carpenter. Shared intentionality. *Developmental Science*, 10(1):121–125, 2007.

[378] F. Warneken, F. Chen, and M. Tomasello. Cooperative activities in young children and chimpanzees. *Child Development*, 77:640–663, 2006.

[379] C. A. Brownell, G. B. Ramani, and S. Zerwas. Becoming a social partner with peers: cooperation and social understanding in one- and two-year-olds. *Child Development*, 77(4):803–821, 2006.

[380] M. Meyer, H. Bekkering, M. Paulus, and S. Hunnius. Joint action coordination in 2½- and 3-year-old children. *Frontiers in Human Neuroscience*, 4(220):1–7, 2010.

[381] J. Ashley and M. Tomasello. Cooperative problem solving and teaching in preschoolers. *Social Development*, 7:143–163, 1998.

[382] T. Falck-Ytter, G. Gredebäck, and C. von Hofsten. Infants predict other people's action goals. *Nat. Neurosci.*, 9(7):878–879, 2006.

[383] F. Heider and M. Simmel. An experimental study of apparent behaviour. *American Journal of Psychology*, 57:243–249, 1944.

[384] S-J. Blakemore and J. Decety. From the perception of action to the understanding of intention. *Nature Reviews Neuroscience*, 2(1):561–567, 2001.

[385] E. Bonchek-Dokow and G. A. Kaminka. Towards computational models of intention detection and intention prediction. *Cognitive Systems Research*, 28:44–79, 2014.

[386] Y. Demiris. Prediction of intent in robotics and multi-agent systems. *Cognitive Processing*, 8:152–158, 2007.

[387] Y. Demiris. Knowing when to assist: developmental issues in lifelong assistive robotics. In *Proceedings of the 31st Annual International Conference of the IEEE Engineering in*

Medicine and Biology Society (EMBC 2009), pages 3357–3360, Minneapolis, Minnesota, USA, 2009.

[388] L. G. Wispé. Positive forms of social behavior: An overview. *Journal of Social Issues*, 28(3):1–19, 1972.

[389] E. Pacherie. The phenomenology of joint action: Self-agency vs. joint-agency. In Axel Seemann, editor, *Joint Attention: New Developments*, pages 343–389. MIT Press, Cambridge MA, 2012.

[390] M. E. Bratman. Shared cooperative activity. *The Philosophical Review*, 101(2):327–341, 1992.

[391] P. R. Cohen and H. J. Levesque. Teamwork. *Nous*, 25(4):487–512, 1991.

[392] S. Butterfill. Joint action and development. *The Philosophical Quarterly*, 62(246):23–47, 2012.

[393] E. Pacherie. Intentional joint agency: shared intention lite. *Synthese*, 190(10):1817–1839, 2013.

[394] P. F. Dominey and F. Warneken. The basis of shared intentions in human and robot cognition. *New Ideas In Psychology*, 29:260–274, 2011.

[395] J. Lindblom. *Embodied Social Cognition*. Cognitive Systems Monographs (COSMOS). Springer, Berlin, (In Press).

[396] J. Piaget. *The construction of reality in the child*. Basic Books, New York, 1954.

[397] L. Vygotsky. *Mind in society: The development of higher psychological processes*. Harvard University Press, Cambridge, MA, 1978.

Index